SpringerBriefs in History of Science and Technology

The *SpringerBriefs in the History of Science and Technology* series addresses, in the broadest sense, the history of man's empirical and theoretical understanding of Nature and Technology, and the processes and people involved in acquiring this understanding. The series provides a forum for shorter works that escape the traditional book model. SpringerBriefs are typically between 50 and 125 pages in length (max. ca. 50.000 words); between the limit of a journal review article and a conventional book.

Authored by science and technology historians and scientists across physics, chemistry, biology, medicine, mathematics, astronomy, technology and related disciplines, the volumes will comprise:

1. Accounts of the development of scientific ideas at any pertinent stage in history: from the earliest observations of Babylonian Astronomers, through the abstract and practical advances of Classical Antiquity, the scientific revolution of the Age of Reason, to the fast-moving progress seen in modern R&D;
2. Biographies, full or partial, of key thinkers and science and technology pioneers;
3. Historical documents such as letters, manuscripts, or reports, together with annotation and analysis;
4. Works addressing social aspects of science and technology history (the role of institutes and societies, the interaction of science and politics, historical and political epistemology);
5. Works in the emerging field of computational history.

The series is aimed at a wide audience of academic scientists and historians, but many of the volumes will also appeal to general readers interested in the evolution of scientific ideas, in the relation between science and technology, and in the role technology shaped our world.

All proposals will be considered.

Lydia Patton · Erik Curiel

Editors

Working Toward Solutions in Fluid Dynamics and Astrophysics

What the Equations Don't Say

 Springer

Editors
Lydia Patton
Department of Philosophy
Virginia Tech
Blacksburg, VA, USA

Erik Curiel
Munich Center for Mathematical
Philosophy
Ludwig-Maximilians-Universität München
München, Bayern, Germany

ISSN 2211-4564   ISSN 2211-4572 (electronic)
SpringerBriefs in History of Science and Technology
ISBN 978-3-031-25685-1   ISBN 978-3-031-25686-8 (eBook)
https://doi.org/10.1007/978-3-031-25686-8

Preface

The editors and authors of this book have been working on aspects of this topic for some time, both separately and in conversation with each other. Erik Curiel's 2010 preprint "On the Formal Consistency of Theory and Experiment, with Applications to Problems in the Initial-Value Formulation of the Partial-Differential Equations of Mathematical Physics" establishes his approach to the partial differential equations of fluid dynamics, including theories of relativistic dissipative fluids. Curiel acknowledges the work of Howard Stein, who did path-breaking research into the question of the relationship between theory and experiment.

Over the past two decades, a significant strain of Susan Sterrett's work has focused on the study of physically similar systems. The themes Sterrett explores in this connection include analogue systems and dimensional analysis.

Colin McLarty is a category theorist who has done substantial research into the history of algebraic geometry, especially the work of Emmy Noether and Alexander Grothendieck.

Jamee Elder's research concerns the methodology and epistemology of large astrophysical experiments, especially those—including LIGO-Virgo-KAGRA and the Event Horizon Telescope—that involve "observing" black holes. This work has encouraged research into the role of equations and models in the interpretation and analysis of observation, including in model validation.

In 2015, Lydia Patton gave a talk called "Fishbones, Wheels, Eyes, and Butterflies" at the MidWest Philosophy of Mathematics seminar at the University of Notre Dame, a longstanding meeting organized by Michael Detlefsen until his passing in 2019, Patricia Blanchette, and Curtis Franks. The 2015 talk was the occasion for Patton, McLarty, and Sterrett to begin a conversation about the mutual interest in the topic of mathematical modeling, the role of equations in physical explanation, and the study of partial differential equations in the philosophy of mathematics and the philosophy of science. That conversation continued in the summer of 2020, in which Susan Sterrett organized a workshop, "What Equations Don't Say", that brought together talks by Patton, McLarty, Sterrett, and Erik Curiel. They were later joined by Jamee Elder, whose work delves into related questions in the philosophy of general relativity and astrophysics. Since the workshop, the authors have continued to discuss and work on their projects, which have culminated in the papers for this volume.

The editors gratefully acknowledge Susan Sterrett's organization of an online workshop in July 2020. The 2020 workshop benefited from the attendance of researchers including Daniele Oriti, George Smith, Bas van Fraassen, Patricia Palacios, Erich Reck, Robert Batterman, Alastair Wilson, Silvia de Toffoli, Germain Rousseaux, Susan Castro, and Brian Hepburn. The authors and editors are very grateful for the workshop participants' contributions to the development of the papers and for the development of conversations on these research topics. Lydia Patton is grateful to Michael Detlefsen, Patricia Blanchette, and Curtis Franks for the invitation to present a paper at the MidWest Philosophy of Mathematics Workshop. Colin McLarty and Susan Sterrett have been galvanizing forces behind the project.

We are grateful to Christopher Wilby, Werner Hermens, and Arumugam Deivasigamani for shepherding the manuscript through the review and publication process.

Blacksburg, USA   Lydia Patton

Contents

Contributors

Jamee Elder Black Hole Initiative, Harvard University, Cambridge, MA, UK

Colin McLarty Case Western Reserve University, Cleveland, Ohio, USA

Lydia Patton Virginia Tech, Blacksburg, VA, USA

Susan G. Sterrett Wichita State University, Wichita, KS, USA

Chapter 1
Introduction

Lydia Patton

Abstract Systems of differential equations are used to describe, model, explain, and predict states of physical systems. Experimental and theoretical branches of physics including general relativity, climate science, and particle physics have differential equations at their center. Direct solutions to differential equations are not available in many domains, which spurs on the use of creative mathematics and simulated solutions.

Keywords Differential equations · Fluid dynamics · Astronomy · Philosophy of science

Philosophical questions arise from the use of differential equations in physical science and in mathematics. In particular, there is a need for sustained attention to the questions in the philosophy of science that arise from the use of differential equations in physical science.[1]

The papers in this volume analyze the use of differential equations in fluid dynamics and astronomy. The central problem at stake is the fact that direct solutions to differential equations are not available in many domains for which the systems of equations are constructed. Lack of a direct or immediate solution means that an equation or system of equations does not have a solution in a domain without employing a method that either restricts the domain, or extends the equations, or both. Mathematicians may refer to the lack of an 'analytic' solution. Or they may refer to the lack of an 'exact' solution. An equation's being without an exact solution can mean a number of things: that no closed form solution can be written, for instance, or that

---

[1] Parker (2017), Morrison (1999), Fillion (2012), Curiel (2010), Batterman (2013), and Mattingly and Warwick (2009) are among those who have contributed philosophical studies of the use of differential equations in the analysis of physical systems.

L. Patton (✉)
Virginia Tech, Blacksburg, VA, USA
e-mail: critique@vt.edu

© The Author(s), under exclusive license to Springer Nature Switzerland AG 2023   1
L. Patton and E. Curiel (eds.), *Working Toward Solutions in Fluid Dynamics and Astrophysics*, SpringerBriefs in History of Science and Technology,
https://doi.org/10.1007/978-3-031-25686-8_1

the available solution does not provide unique exact values for variables of interest, but provide approximate or perturbative solutions instead.[2]

The papers that follow focus on the Navier-Stokes equations in fluid dynamics and Einstein's field equations of general relativity as they are employed in astrophysics. The Navier-Stokes equations do not have immediate exact solutions for many physical fluid systems. Similarly, the Einstein field equations of general relativity do not have direct solutions in many domains of interest, including the merger of astronomical bodies, which emanates gravitational waves.

And yet, scientists and engineers work with these equations every day. Ingenious methods have been devised, which may involve finding simulated solutions for artificial or idealized situations and then extending those solutions to actual cases; or finding 'weak' solutions which may not have derivatives (solutions) everywhere, but which nonetheless satisfy the equations in some restricted sense; or finding another strategy to extend the equations to the domain of interest. In many cases, solutions that initially apply only in restricted or idealized contexts are extended to a wider set of physically realistic situations.

Equations may be used in scientific reasoning to yield predictions and explanations: to predict states of a system and uniquely specify the values of variables. Equations can play these roles directly when immediate solutions are available. This collection goes beyond that familiar case to explore what happens when direct solutions are absent, or when the task at hand is precisely to find a way to connect the equations to a specific physical situation.

One of the most striking results learned early on in real analysis is the fact that there are many more irrationals than rational numbers. Early education often presents irrational numbers as exceptions, so it comes as a surprise to find that they are more numerous than the rational numbers. The application of differential equations to physical contexts reveals a similar situation. Physical reasoning using differential equations usually is presented as follows: scientists select an appropriate equation or system of equations, find a solution, and thereby determine the evolution and properties of a physical system. But cases in which a direct solution is not available and scientists must work out a solution, or work in the absence of a solution, are far from the exception. Working with equations can involve reasoning toward, from, and around equations as well, in the absence of a solution in the domain of interest.

The papers focus on how scientists reason with, around, and toward differential equations in fluid dynamics and astrophysics in the absence of immediate solutions. The process of reasoning may involve extending available solutions to differential equations to initially inaccessible domains.[3] Or, it may involve finding novel ways of simulating or modeling the domain so that it can make contact with the equations, or new methods for calculating, or new ways of measuring. McLarty 2023 focuses

---

[2] These are not the only definitions in common use. See Fillion and Bangu 2015 and McLarty 2023 for discussion of types of solutions.

[3] A conversation with Hasok Chang put a clear focus on 'stretching' or 'extending' equations, which is a helpful way of describing one aspect of these situations.

serious attention on how methods of calculation can make a difference in methods for solving, but also for reasoning more broadly with, differential equations.[4]

A target physical phenomenon, say turbulent fluids or merging black holes, must be characterized in a certain way to be tractable using differential equations in the first place. One must choose a way to represent and measure physical variables. Beyond that, scientists must determine ways to represent a physical domain as a system, including how to determine initial conditions, boundary conditions, and constraints on system evolution.[5] In many cases, establishing initial and boundary conditions and kinematical constraints to characterize a given problem allows for differential equations to be applied to solve that problem.

For this collection, Sterrett 2023 builds on her earlier work on symbolic reasoning, physical analogy, dimensional analysis, and modeling. She argues here that there is a stronger and broader role for mathematics in characterizing physical systems than in solving equations that are already known. Sterrett, McLarty 2023, and Patton 2023 urge that we evaluate the mathematics used in reasoning about physical systems in a more flexible, creative way—including the reasoning used to characterize and to measure those systems.

Setting up a solution to differential equations in a specific physical domain requires finding precise ways to determine the conditions and constraints under which a solution is possible in that domain. Even in highly theoretical fields of physics, setting up a problem involves deep understanding of the physical situation at hand, as Elder's and Sterrett's papers for this collection show brilliantly. Abstract theoretical research may require increasingly precise understanding of a physical system, because counterfactual reasoning is based on knowing the exact parameters and constraints to alter when departing from the concrete properties of the system. Patton 2023 argues that weak and simulated solutions to equations can allow for heuristic extension of structural, physical reasoning into situations where the equations lack direct solutions.

On the other side, reasoning about the physical properties of a system may require significant theoretical or modelling resources. For instance, as Elder 2023 explains, the data gathered by the new generation of gravitational wave detectors is not independently informative. Data is filtered through a bank of waveform templates generated using both empirical and theoretical reasoning, after which post-data analysis allows for estimates of the physical parameters of the target system.[6] The numerical simulations and waveform templates used by the LIGO-Virgo-Kagra Scientific

---

[4] McLarty draws an analogy between the Navier-Stokes equations and Emmy Noether's theorems on this score.

[5] Curiel has argued that "it is satisfaction of the kinematical constraints—fixed, unchanging relations of constraint among the possible values of a system's physical quantities—that ground the idea of the individual state of a system as represented by a given theory. If the individual quantities a theory attributes to a system do not stand in the minimal relations to each other required by the theory, then the idea of a state as representing that kind of system cannot be cogently formulated, and without the idea of an individual state of a system one can do nothing in the theory to try to represent the system" (Curiel 2016).

[6] Elder 2023 and Patton 2020 provide details, including references to the scientific literature on this topic.

Collaboration are constructed and validated using vast theoretical and empirical resources including probabilistic methods, methods of approximation, dynamical equations and reasoning, and empirical input. Elder 2023 shows that the parameters, dynamical reasoning, and empirical information crystallized in the waveform templates supports their extraordinary flexibility in application.

An essential tenet of this volume is that a great deal of creative mathematical reasoning can go on in physics without direct solutions to the equations in the field of interest. Our aim is to integrate this fact, known to the scientists themselves and to historians of mathematics for some time[7], into the philosophical analysis of physical reasoning.

Lydia Patton

References

Batterman R (2013) The tyranny of scales. In the Oxford handbook of philosophy of physics. Oxford University Press, Oxford, pp 255–86

Curiel E (2010) On the formal consistency of theory and experiment, with applications to problems in the Initial-Value Formulation of the Partial-differential equations of mathematical physics. Preprint. http://philsci-archive.pitt.edu/8660/

Curiel E (2016) Kinematics, dynamics, and the structure of physical theory. Preprint. https://doi.org/10.48550/arXiv.1603.02999

Fillion N (2012) The reasonable effectiveness of mathematics in the natural sciences. Dissertation, The University of Western Ontario

Gray J (2021) Change and variations: a history of differential equations to 1900. Springer, Dordrecht

Mattingly J, Warwick W (2009) Projectible predicates in analogue and simulated systems. Synthese 8(3):465–482

Morrison, M (1999) Models as autonomous agents. In Morgan and Morrison (eds.) Models as mediators. Cambridge University Press, Cambridge, pp 38–65

Parker W (2017) Computer simulation, measurement, and data assimilation. Br J Philosop Sci 68(1):273–304

---

[7] A recent, much-anticipated history of differential equations is (Gray 2021); this and much of Gray's earlier work, including his work on Poincaré, explores the development of equations in satisfying detail.

Chapter 2
How Mathematics Figures Differently in Exact Solutions, Simulations, and Physical Models

Susan G. Sterrett

Abstract The role of mathematics in scientific practice is too readily relegated to that of formulating equations that model or describe what is being investigated, and then finding solutions to those equations. I survey the role of mathematics in: 1. Exact solutions of differential equations, especially conformal mapping; and 2. Simulations of solutions to differential equations via numerical methods and via agent-based models; and 3. The use of experimental models to solve equations (a) via physical analogies based on similarity of the form of the equations, such as Prandtl's soap-film method, and (b) the method of physically similar systems. Two major themes emerge: First, the role of mathematics in science is not well described by deduction from axioms, although it generally involves deductive reasoning. Creative leaps, the integration of experimental or observational evidence, synthesis of ideas from different areas of mathematics, and insight regarding analogous forms are required to find solutions to equations. Second, methods that involve mappings or transformations are in use in disparate contexts, from the purely mathematical context of conformal mapping where it is mathematical objects that are mapped, to the use of concrete physical experimental models, where one concrete thing is shown to correspond to another.

Keywords Equations · Models · Mathematics · Conformal mapping · Physically similar systems · Simulations

2.1 Introduction

The role of mathematics in scientific practice is too readily relegated to that of formulating equations that model or describe what is being investigated, and then finding solutions to those equations. That is a tidy but incomplete account of the role of mathematics in science. For one thing, mathematics is involved in experimental

S. G. Sterrett (✉)
Wichita State University, Wichita, KS, USA
e-mail: susan.sterrett@wichita.edu

© The Author(s), under exclusive license to Springer Nature Switzerland AG 2023
L. Patton and E. Curiel (eds.), *Working Toward Solutions in Fluid Dynamics and Astrophysics*, SpringerBriefs in History of Science and Technology,
https://doi.org/10.1007/978-3-031-25686-8_2

investigations to help answer investigative questions in ways other than finding solutions to mathematical equations. Further, the equations themselves can be involved in scientific practice in a variety of ways, some of which are substantively different from others. Identifying, clarifying, and explaining those different ways is my topic in this paper. To clarify the differences, I'll focus my discussion in this short paper on comparing how mathematics is involved in exact solutions, simulations, and experimental physical models.

2.2   Exact Solutions of (Differential) Equations

Which mathematical methods are used in solving equations of mathematical physics depends, of course, on the kind of equation. If we are talking about differential equations, then what is meant by a solution to the equation is a function that satisfies the equation. Sometimes the problem also specifies boundary values or initial values, adding more specific requirements that the function proposed as a solution to the differential equation must meet. Though it is possible to show, for some classes of equations, the existence and uniqueness of a solution, this is not the case in general. The Navier-Stokes equations in fluid dynamics are a set of partial differential equations, and the question of whether smooth, physically reasonable solutions to them exist is one of the Clay Mathematics Institute's Millenium problems.[1]

The study of solutions of partial differential equations, i.e., differential equations that involve partial derivatives, is an entire field of study unto itself. Partial differential equations are further classified into a variety of major types. The details of the classification are too involved to lay out here. For this discussion, I mention a few specific differential equations that have been given special names: the wave equation, the diffusion equation, and Laplace's equation. Many more could be mentioned.

Although the wave equation, the diffusion equation, and Laplace's equation are special cases of a differential equation, each constitutes an entire area of research in mathematical physics. Each applies to an indeterminately wide variety of phenomena in physics. The wave equation applies to electrical waves as well as mechanical waves and many other kinds of waves; the diffusion equation to heat diffusion (conduction), diffusion of particles, and diverse kinds of diffusion; and Laplace's equation likewise applies to a wide range of phenomena that arise in the study of heat, fluid flow, electrostatics, and similar phenomena in physics.

Our question in this context is thus how mathematics is involved in finding exact solutions of partial differential equations in mathematical physics. A straightforward approach to answering it is immediately thwarted, though, as there is no general method for finding exact solutions to partial differential equations. Deriving solutions to partial differential equations is not so much a matter of deductive logic, much less symbolic logic and set theory, as it is a matter of creativity by someone knowledgeable

---

[1] The official problem description is here: "Existence and Smoothness of the Navier-Stokes Equation" by Charles R Fefferman (https://www.claymath.org/sites/default/files/navierstokes.pdf).

in disparate fields of mathematics, some of which might be far afield from the equation in question. Often a significant amount of intellectual work is involved in identifying and reflecting on features of the equations to be solved, on boundary conditions, and on the difference that various kinds of boundary conditions make to the nature and existence of solutions, symmetries of a particular problem, and so on. Then, it is often a matter of resourcefulness. Even though I will later want to emphasize that there are other methods in mathematical physics than finding exact solutions, I still wish to register an appropriate appreciation of what is involved in finding exact solutions.

One of the most elegant and beautiful methods used in finding a function that is an exact solution to a partial differential equation is conformal mapping. The basic idea of the method is familiar from cartography: to map figures from the spherical earth onto a flat surface while at the same time preserving angles locally involves stretching and translation of the figures on the surface of a globe. The geometrical problem of which direction to head in to get from one point to another on the globe is solved by finding the line between those two points on the flat map, even though distances and areas are not correct on the flat map. In complex analysis, this basic idea was applied to map figures and graphs from one flat two-dimensional surface to another via transformations that involve stretching/shrinking and translation. Riemann's 1851 dissertation is credited with this creative suggestion (in spite of details about the proof and distinctions about the use of the term 'conformal' that would be brought up in discussing his formulation and proof today.) Ullrich notes:

> As an application of his [Riemann's] approach he gave a 'worked-out example', showing that two simply connected plane surfaces can always be made to correspond in such a way that each point of one corresponds continuously with its image in the other, and so that corresponding parts are 'similar in the small', or conformal … (Ullrich 2005, 454)

Conformal mapping was developed and used in complex analysis to find exact solutions to partial differential equations with great success in the 19th century. Bazant[2] remarks that

> The classical application of conformal mapping is to solve Laplace's equation:
>
> $$\nabla^2 \varphi = 0$$
>
> i.e. to determine harmonic functions in complicated planar domains by mapping to simple domains. The method relies on the conformal invariance of Eq. (1.1) [above], which remains the same after a conformal change of variables. (Bazant 2004, 1433)

Thus, problems involving complicated shapes of objects, boundaries, or surfaces would first be mapped, by a savvy choice of change of variables, to a new domain in which the shapes made the problem tractable (e.g., mapping an airfoil to a circle). Then, after solution in the new domain, the solved problem would be transformed back to the original domain, using the inverse of the function that had been used to map the problem from the original domain in which the shape was complicated to the

---

[2] I am indebted to Lydia Patton for introducing me to the work of Martin Z. Bazant, in her talk "Fishbones, Wheels, Eyes, and Butterflies", given at the Midwest Philosophy of Mathematics Workshop 17, University of Notre Dame, 12–13 November 2016.

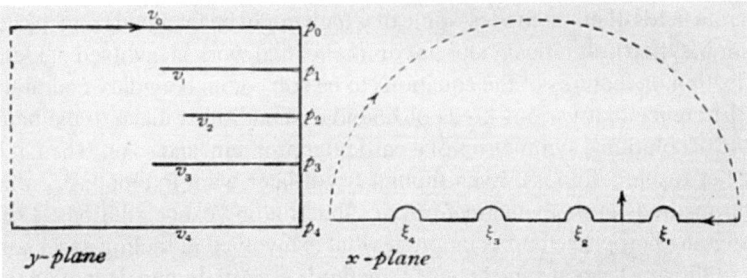

Fig. 2.1 Schwarz-Cristoffel transformation of upper half-plane

domain in which the problem was tractable. Using the inverse function mapped the (now solved) problem from the domain in which it was solvable back to the original domain. This meant that the solution to the problem was mapped back to the original domain, too. So, it was a way of obtaining solutions to mathematical problems that were intractable as originally stated.

As mathematically elegant as this means of solving intractable problems was, it was not without a concrete precedent. Navigators in the 16th century could answer the question of which direction they should head on the spherical surface of the actual earth by consulting a flat map in which directions between points on the earth were preserved between corresponding points on the flat map—in spite of how distorted land shapes and area proportions were on the flat map. Likewise, a scientist could now solve problems that were otherwise intractable by working on the suitably transformed problem instead. The hard work is in finding the transformation that transforms the problem in such a suitable manner that the transformed problem is tractable and allows these two-way mappings. It is crucial to obtaining exact solutions to sets of differential equations via conformal mapping to be able to identify an appropriate transformation, i.e., an appropriate mapping function. Later in this paper, we will see other contexts in which attending to the role that an appropriate mapping function plays is likewise crucial, and why its role is so philosophically significant.

One of the earliest, and still well-known, transformations is the Schwarz-Cristoffel transformation. An example of the many problems to which it has been applied, taken from a nineteenth century text, is shown below.[3] While the figures are simple, it takes some time to contemplate and fully appreciate what the mapping effects, e.g., to understand just what the interior points of the polygon in Fig. 2.2 are mapped to.

The same mapping can be used to map all sorts of geometrical figures occurring in a variety of fields of physics. Sometimes a conformal transformation maps points at infinity to points on a figure. In fact, the Schwarz-Cristoffel style of map mentioned above is often illustrated by showing how the infinite (upper) half-plane can be mapped to various objects, such as a triangle. The visual insight as described in the 19th century text uses both Figs. 2.1 and 2.2 above: "when x turns to the right into

---

[3] Figures 2.1 and 2.2 are taken from Harkness and Morley 1898, 321).

Fig. 2.2 Schwarz-Cristoffel
transformation: polygon

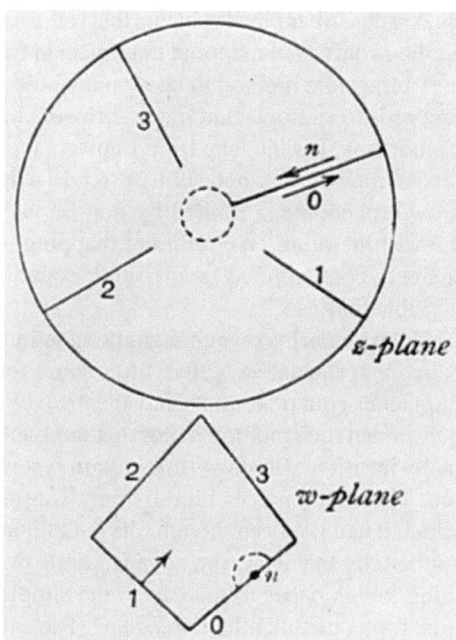

the upper half-plane, as indicated by the arrow in the right-hand part of Fig. 2.1, w turns also to the right into the interior of the polygon, as indicated by the arrow in the w-polygon of Fig. 2.2. Hence the upper half of the x-plane maps into the interior of the w-polygon" (Harkness and Morley 1898, 322).

Bazant comments that "important analytical solutions were thus obtained for electric fields in capacitors, thermal fluxes around pipes, inviscid flows past airfoils, etc." citing twentieth-century works (Bazant 2004, 1434). Today there are computer programs that can carry out the transformations. These applications developed from the methods first that first arose in the mid-nineteenth century. However, the story of that method did not end once it achieved success in finding exact solutions to problems expressible in the form of Laplace's equation. In the early twenty-first century, the creativity and resourcefulness of some mathematicians led them to consider how this method of solution might be used elsewhere, i.e., for a different kind of problem in mathematical analysis. Though conformal mapping developed from within the study of Laplacian problems, it did not stay there. Bazant exuberantly reported in 2004 that:

> Currently in physics, a veritable renaissance in conformal mapping is centering around 'Laplacian-growth' phenomena, in which the motion of a free boundary is determined by the normal derivative of a harmonic function. ... Such problems can be elegantly formulated in terms of time-dependent conformal maps, which generate the moving boundary from its initial position. (Bazant 2004, 1434)

Bazant describes developments that followed using iterated conformal maps, and then applied conformal mapping to Laplacian fractal-growth phenomena. His 2004 paper then brings the method to bear on a whole different class of problems in analysis: it proceeds from Laplacian fractal-growth phenomena to non-Laplacian fractal growth phenomena. It's the leap from Laplacian to non-Laplacian that's notable here: "One of our motivations here is to extend such powerful analytical methods to fractal growth phenomena limited by non-Laplacian transport processes. Compared with the vast literature on conformal mapping for Laplace's equation, the technique has scarcely been applied to any other equations." In fact, it just hadn't been thought possible to do so.

He explains his own thought process in applying the technique beyond Laplacian equations. Bazant says that, after some reflections on the fundamental features of Laplacian equations that lend themselves to solutions by conformal mapping, he questioned the common belief that the Laplacian operator is unique in being conformally invariant. He shows that certain systems of equations are conformally invariant, too. This then enables him to apply conformal mapping techniques to problems to which it had not been thought the technique of conformal mapping could be applied. As he tells the story, the advance boils down to one key insight, namely: "Everything in this paper follows from the simple observation that the advection operator transforms just like the Laplacian" (Bazant 2004, 1436).

Obviously coming up with these mathematical methods involves much more than deductive logic. What else? Analogy, knowledge of a variety of areas of mathematics, identifying and questioning a presumption prevailing among mathematicians, and creative leaps. So, it took more than the straightforward application of deduction in mathematics to come up with the methods Bazant reveals.

Once these methods are in hand, though, we can ask where and how mathematics is involved in the activity of using these methods to find exact solutions. Here we can be more precise about an answer: Not only is mathematics involved in putting the problem from physics into the mathematical language of differential equations, but it is involved in transforming that mathematical problem into one that is tractable. Once the problem is transformed, it can be solved in its novel form, and then one can use the inverse of the transformation process to put it back into the original domain, where it can be interpreted in the language of physics. Hence the mathematical functions that effect the mappings, and the construction of the domains where problems can be made tractable, are of crucial significance.

2.3   Simulations

Exact solutions of (partial differential) equations in mathematical physics are the exception rather than the rule, though.[4] Hence calculational methods have been developed. Numerical methods predate the current high-speed computational devices and

---

[4] This point is also made and discussed in McLarty, this volume Colin (2023).

systems in use today. It is often pointed out that some numerical methods associated with the calculus are due to, and hence in use as early as, Isaac Newton, but numerical methods are at least as old as papyri. New methods continue to be developed and enjoy widespread use, such as the finite element method in the latter twentieth century and agent-based modeling (cellular automata) in the twenty-first century, which will be discussed separately below. The widespread use of numerical methods and simulations does not, however, render exact solutions useless or unnecessary—when an exact solution does exist for certain cases, the exact solution serves the important role of a 'benchmark' against which numerical methods or simulations are compared.[5]

2.3.1 Simulations of (Differential) Equations of Mathematical Physics

The simulations discussed here are numerical or other kinds of formal methods used in mathematical physics to yield approximations of solutions to the equations of mathematical physics. Whereas the methods used in finding exact solutions are elegant and deeply satisfying intellectually, the methods used in finding good simulations are nifty and deeply satisfying as well, in terms of their ability to provide practical and useful answers to analytically intractable problems. A wide variety of numerical methods have been developed, and the way that mathematics is involved in them varies accordingly. Some revolve around finite difference methods whereas others rely on probability, such as the Monte Carlo method. Perturbation theory and theory of errors are foundational theories for other numerical methods. Yet, with respect to the question in this paper, some generalizations can be made.

The ways that mathematics may be involved in the simulations under discussion here can be categorized in terms of three kinds of activity.

First, the development of equations, algorithms, and other formal methods in order to turn a problem involving the differential equation or systems of them into a problem that is well suited for computation.

Second, verification that the methods so developed will produce results that are solutions of the problem, within a certain band of error, and under certain limitations on the range of the variables in the problem. This step usually involves mathematical proofs and deductive reasoning, and often identifies the range of variable values over which the method is to be used.

Third, validation of the simulation via comparison with either exact solutions (used as 'benchmarks'), observational data, or experimental data, such as the results of an experimental physical model.

To be clear, these activities are not performed on an individual basis by an individual researcher very often anymore. Not only the first step of transforming the problem into one that is computationally tractable, but the steps of verification and validation,

---

[5] The use of benchmarks in simulations is likewise discussed in Patton (2023).

are often carried out by communities of researchers (including researchers working for vendors of software) and inherited by subsequent researchers in the form of software. Hence it is quite common for someone to design and run a simulation without performing any of the three steps themselves. Ideally, such users would understand at least the basics of the first step (i.e., transforming the problem into one that is computationally tractable), and the second step, of understanding the ranges outside of which the method has not been verified. For, the users of the community-developed and verified software must make judgments in choosing which software to use for the problem they wish to solve, and in implementing the computationally tractable version of the problem in the software. Neither of these decisions is trivial, nor, if done well, free of mathematical reasoning. The third step, validation of the simulation, involves making comparisons between the values of quantities calculated by the simulation and values obtained some other way; either using an analytically closed form solution (exact solution) in the benchmark case, observational data, or experimental data from a specially constructed experimental physical model. Here the usual role that mathematics plays in working with measurements and uncertainties is involved.[6]

The topic of this paper is the role of mathematics in various methods in mathematical physics. However, the inclusion of the validation step hints at something else that is noteworthy. Computer software such as software that integrates fluid flow, heat conduction, and mass transfer in a computational flow dynamics program is not totally a matter of mathematics, even when numerical methods are included among mathematical methods. For empirical results of experimental studies are involved in the first step (development of formal methods that appropriately include physical factors), though sometimes in hidden ways (e.g., judgments as to whether a certain factor can be neglected), as well as in the third step (validation of the method/algorithm/software). If understood as a single linear three-step process, however, even if this point is appreciated, the description still does not quite reveal the extent to which experimental results are involved in the computation, for the process of building simulations involves much trial and error, iteration, and feedback—even in mathematical physics.

2.3.2   Agent-Based Simulations

The way that mathematics is involved in agent-based simulations is distinctive enough that they deserve separate mention. These kinds of simulations aren't implementations of algorithms to obtain approximations to solutions of equations. In agent-based simulations, rules are devised to proscribe the behavior of many individual agents acting in the same environment, in an attempt to model complex systems in which it is patterns of behavior that are of interest. Usually these rules for agents

---

[6] Relevant philosophical work on the topic of numerical methods and/or simulations can be found in Fillion (2017) and Lenhard (2019).

are based in part on facts about other agents, such as how many agents of a certain kind are left, what they are doing, or how close other agents are to it. However, the rules are usually fairly simple rules mathematically speaking. So is the behavior of an individual agent: either ON or OFF (or 'alive' or 'dead'). Different patterns arise depending on the initial configuration, and one soon begins to see the patterns of agent behavior as objects in their own right. They come to be regarded as agents and actions at a higher level than individual rule-following agents. Most philosophers were introduced to these ideas via the "Game of Life" by John H. Conway, often via Daniel Dennett's 1991 "Real Patterns" (Dennett 1991, 27–51).

Agent-based simulations today are much more sophisticated. For instance, the environment in which the agents act can include resources that influence the agents' capabilities to act, and the algorithms contain parameters that amplify or dampen rates or intensities. Swarm behavior of birds and fish, as well as the behavior of crowds and traffic, have been modeled with such agents. The kinds of uses researchers have made of one such simulation program alone—Uri Wilensky's NETLOGO—is seemingly unlimited: art, biology, epidemiology, earth science, chemistry, hydrology, political science and social science, and so on. Hundreds of thousands of people from many different disciplines have used it.[7] Conway's much simpler "Game of Life" can be programmed as a NETLOGO model, too (Wilensky 1998 and Wilensky 1999).

Since some phenomena that are described by differential equations, such as diffusion of particles and predator-prey interactions, can be modelled using agent-based models as well, the question of how the mathematics used in each are related naturally arises. NETLOGO has been expanded beyond agent-based models, to include a "Systems Dynamics" modeler, so that both the agent-based approach and the kind of approach used to develop differential equations to describe the same behavior can be taken. Wilensky describes the difference in how Wolf-Sheep predation is modeled when using the Systems Dynamics Modeler, versus the NETLOGO agent-based simulation modeler, as follows.

> System Dynamics is a type of modeling where you try to understand how things relate to one another. It is a little different from the agent-based approach we normally use in NetLogo models. With the agent-based approach we usually use in NetLogo, you program the behavior of individual agents and watch what emerges from their interaction. In a model of Wolf-Sheep Predation, for example, you provide rules for how wolves, sheep and grass interact with each other. When you run the simulation, you watch the emergent aggregate-level behavior: for example, how the populations of wolves and sheep change over time. With the System Dynamics Modeler, you don't program the behavior of individual agents. Instead, you program how populations of agents behave as a whole.[8]

The way mathematics and logic are involved in the agent-based simulation, then, is really not just "a little" different from the way it is involved in the System Dynamics Modeler; it is strikingly and fundamentally different. For there is no equation describing the dynamics of the populations of predator and prey involved. Rather,

---

[7] A running list of publications in which NetLogo was used or mentioned, many in scientific journals, is maintained on NetLogo's website. The list contains hundreds if not thousands of papers from 1999 to the present, and more are added daily.

[8] "NetLogo Systems Dynamic Guide", n.p.

rules for individuals in the population are formulated, and in running the simulation, the dynamics of the predator and prey populations "emerges" from the agents acting according to those rules, as a matter of logical deduction.

However, models in use today are not always one or the other (i.e., not completely agent-based nor completely equation based); some are hybrid. NETLOGO is often integrated into other models, as part of a more comprehensive program. For instance, in hydrology, one model in use combines agent-based approaches in NETLOGO for the effect of agents who use water, along with finite-difference methods for solving equations of hydrological models of water flow (Castilla-Rho 2015). Further complicating any attempt at a strict taxonomy are newer developments in which the agents' behavior is continuous rather than discrete, as in the agent-based programs developed to model behavior of continua. The Turbulence model is one example (Wilensky 2003).; another is the vibration of a plate or membrane (Wilensky 1997).

When using agent-based models to investigate what emerges from agent-based approaches, apart from the aim of solving differential equations in mathematical physics, it could be that very little mathematics is involved, even when the behavior that emerges turns out to be a numerical or approximate solution to a differential equation. But in such a case, one is not looking for a solution to an equation, and thus there is no right or wrong in the matter. Inasmuch as the models are used to solve problems of mathematical physics, the process is broadly the same as the three part process described above for numerical methods: development of the formal method, verification of the formal method, and (depending on the aim of the modeler) validation of the formal method. Thus there is more involved than mathematics: observation and experimentation are involved, too.

2.4 Role of Mathematics in Experimental Physical Models

As mentioned earlier, although exact solutions have been found for a few special cases of the Navier-Stokes equations and other systems of partial differential equations, no general method for their solution is known. Conformal mapping is an elegant and powerful method, but it is not a general method; it relies on a researcher's creativity and resourcefulness. Observed phenomena related to turbulence, such as its onset and the separation of the laminar from turbulent regions of flow, are still not well understood nor mathematically tractable for many configurations. But methods of similarity in physics, which were used by Galileo and Newton in mechanics and dynamics (and likely before by others), were developed further for hydrodynamics in the centuries that followed. An examination of the reasoning used in problems in physics from pendula to vibration of plates, if traced back to their sources, would reveal the importance of using similarity along with observations and/or physical models to inform both the formulation of the differential equations and the methods of solving them.

Numerical methods and simulations are generally more cost-effective than building experimental physical models and setups for each and every situation one wishes to investigate, but the widespread use of numerical methods and simulations does not mean that they could be employed independently of the information gained from experimental physical models and setups. As there seems to be cultural amnesia about the significance of the role of experimental physical models and physically similar systems, at least among philosophers of science[9] and philosophers of mathematics, I ask the reader's indulgence here as I take the time to describe some early history of the topic that will be helpful in understanding the philosophical points in this paper.[10]

2.4.1   Physical Analogies in the Early Twentieth Century

In his "Exact Solutions and Physical Analogies for Unidirectional Flows" Bazant (2016) notes that "In contrast to the more familiar case of Laplace's equation . . . conformal mapping cannot be as easily applied to Poisson's equation, since it is not conformally invariant." He notes, however, that "mathematical insights allowed Poisson's equation to be solved experimentally, long before it could be solved numerically on a computer." What can it mean to say that an equation can be solved experimentally? What kind of 'mathematical insights' could enable that?

The 'mathematical insights' Bazant names are "mathematical equivalence of beam torsion and pipe flow . . . [and] convective heat transfer"; and analogies with elastic membrane deflections, soap bubbles, and "the potential profile of electrically conducting sheets." From such insights, scientists were able to build a setup of one kind to determine the behavior—and for specific cases, determine values of quantities—of another kind. One of the earliest, most well-known, and tractable of these was the use of the analogy from membranes. A membrane was easy to create from soap film, hence it became known as "Prandtl's soap-film analogy." Prandtl's insight was that an analogy between two quite different phenomena could be made, "which could be described by the same differential equation if . . . specific parameters were replaced in each case by other [specific parameters.]" In Eckert's biography of Prandtl, he writes about Prandtl's paper describing analogous physical setups: "In the first case, the distortion of a soap membrane which is stretched over the opening of a container and bulges outward as a result of a small positive pressure in the container is considered; in the other, the twisting (torsion) of a bar that has the same diameter as the opening of the container" (Eckert 2019, 29). Eckert goes on to detail how the same differential equation describes two different kinds of quantities in the two quite different setups: "In the first case, the differential equation describes the lateral buckling as a result of the positive pressure in the container; in the second case,

---

[9] Sherrilyn Roush is a rare exception. In "The Epistemic Superiority of Experiment to Simulation" (Roush 2018) she recognizes that "the solver" in a computer simulation incorporates sources other than 'the theory' (p. 4886).

[10] A longer treatment is given in Sterrett, Susan G. "Physically Similar Systems: a history of the concept" (Sterrett 2017).

the tension along the circumference of a bar cross-section induced by the twisting (torsion moment) of the bar" (Eckert 2019, 29).

Now, how, exactly, does one use soap film to get the solution of a problem using a physical analogy? Here's how: You would only have to construct the setup with the soap membrane stretched over the opening of a container. Then, you could take measurements as follows: "The angle of slope of the bulging membrane in the first case corresponds to the shearing stress on the cross-sectional outline of the bar in the second case. The volume over the opening caused by the bulging of the soap membrane corresponds to the torsional stiffness of the bar" (Eckert 2019, 29). This is how problems can be solved by measurements of the membrane in the setup of the soap film membrane—i.e., solved 'experimentally'. But how is a set of measurements of distance in the soap film informative about stress in a bar?

Here the form of the mathematical (partial) differential equations displays the analogy.

For the bar: [T]he torsion of a bar along its long axis (x-axis in a cartesian coordinate system) is described by a stress function $y(y, z)$, . . . The stress function conforms to the equation

$$\frac{\partial^2 \psi}{\partial y^2} + \frac{\partial^2 \psi}{\partial z^2} = 2G\theta$$

where G is a material constant (torsion modulus) and θ the torsion per unit of length.

For the (soap-film) membrane: A membrane that is stretched in the yz plane over an opening corresponding to the bar cross-section (tension S) and subjected on one side to a constant pressure p will bend towards the other side by an elongation $u(y, z)$. This elongation is described by the equation

$$\frac{\partial^2 u}{\partial y^2} + \frac{\partial^2 u}{\partial z^2} = \frac{p}{S}$$

From these two equations, *an analogy is produced between the stress function for the torsion of a bar and the bulging convexity of a membrane over an opening of the same surface as the bar cross-section*[11]:

$$\psi = \frac{2G\theta S}{p} u$$

You cannot see or easily directly measure the stress in the bar but you can see—and measure—the distance that the soap membrane is bulging. So you construct the soap film setup, measure the bulge u, and from it and the equation $\psi = \frac{2G\theta S}{p} u$ compute the stress in the bar.

The insight arises from the mathematical form, i.e., the analogy can be intuited because the physical quantities in both the bar and the membrane are expressed in terms of functions that are solutions to partial differential equations—not due to any experiential familiarity or knowledge about torsional stress in bars or deflections in thin membranes. Prandtl felt the approach could be used in many more fields. To facilitate recognition of analogous physical situations by looking at the mathematical equations, he felt it was of utmost importance to standardize mathematical

---

[11] Eckert (2019), 29, emphasis added.

expressions across different scientific and technical areas of science and engineering. The idea was that doing so would increase the opportunities for the kind of mathematical insights he had with the soap membrane analogy. The reason it was so important to facilitate such insights was that seeing such analogies would allow people to obtain solutions to partial differential equations not available any other way. I take it that this is just what Bazant was referring to in saying that "mathematical insights allowed Poisson's equation to be solved experimentally, long before it could be solved numerically on a computer" (Bazant 2016, 024001–2).

Bazant's 2016 paper on physical analogies published in *Physical Review Fluids* shows how really fruitful this approach is, even today. He writes about the "common mathematical problem [that] involves Poisson's equation from electrostatics

$$-\nabla^2 u = k$$

typically with constant forcing k and Dirichlet (no-slip) boundary conditions on a two-dimensional domain", and notes that "The same problem arises in solid mechanics for beam torsion and bending" and, in fact, in two dimensions, arises "for a remarkable variety of physical phenomena" (Bazant 2016, 024001–2). He provides a survey and then expands the number of physical analogies even farther. He lists a total of seventeen, sketched in a figure; the caption lists them as follows:

Seventeen analogous physical phenomena from six broad fields, all described by Poisson's equation in two dimensions. Fluid mechanics: (a) Poiseuille flow in a pipe, (b) circulating flow in a tube of constant vorticity, and (c) groundwater flow fed by precipitation. Solid mechanics: (d) torsion or (e) bending of an elastic beam, and (f) deflection of a membrane, meniscus, or soap bubble. Heat and mass transfer: (g) resistive heating of an electrical wire, (h) viscous dissipation in pipe flow, and (i) reaction-diffusion process in a catalyst rod. Stochastic processes: (j) first-passage time in two dimensions, (k) the chain length profile of a grafted polymer in a tube, and (l) the mean rate of a diffusion-controlled reaction. Electromagnetism: (m) vector potential for magnetic induction in a shielded electrical wire, and the electrostatic potential in (n) a charged cylinder or (o) a conducting sheet or porous electrode. Electrokinetic phenomena: (p) electro-osmotic flow and (q) streaming current in a pore or nanochannel.

One of them is especially surprising, and shows the creativity and intellectual insight in recognizing these analogies: the stochastic processes (j, k, and l).

The role mathematics plays here is quite explicit: first, an understanding of analysis as used in mathematical physics allows someone to formulate partial differential equations in a canonical form; second, comparison of partial differential equations from various parts of mathematical physics provides opportunities to recognize analogies between very different areas of mathematical physics; and third, once an analogy is recognized, the equation permits someone to use a setup analogous to the one that one wishes to have a solution of, to obtain a solution. The mathematical equation also provides the mapping from a measured quantity in one of the setups to the inferred quantity in the other.

There are philosophers of science who are familiar with this kind of physical analogy, in a general way. A common example cited is the harmonic oscillator, which is the linear differential equation for a mass on a spring—and for many other physical systems in nature, as well. Francisco Guala and Chris Pincock's works, to take two recent examples, exhibit familiarity with cases where an equation is instantiated by more than one situation, and Pincock specifically mentions partial differential equations, including Laplace's equation and Poisson's equation. Pincock briefly discusses cases of scale similarity and dynamical similarity (experimental scale models); he assumes that the dimensionless parameters used to effect this kind of similarity are obtained from the equations that describe the two similar situations.[12] We shall see that although this may often be so, it is not necessarily the case: there is another basis for similarity that does not require knowledge of even the governing equations. The explanation for how we can establish that kind of similarity without knowledge of the governing equation requires looking more broadly than either of these two philosophers have, into the foundations of metrology and the relationship between different kinds of mathematical equations and how they are related to physical quantities. We turn now to that method: the method of physically similar systems, via the use of dimensional equations.

2.4.2   The Method of Physically Similar Systems

The use of similarity by European scientists in the nineteenth and early twentieth century was wide-ranging if not ubiquitous. There is not room here to convey the breadth and depth of the uses made of similarity in physics, but I have tried to do so elsewhere, and I refer the interested reader to that paper, "Physically Similar Systems: A history of the concept" (Sterrett 2017). For the question that is the focus of this paper, the role of mathematics in experimental physical models, I pick out a few exceptional papers to highlight the distinctiveness of the method.

Against a backdrop of the impressive and exciting accomplishments made by reasoning from analogy, resulting from the ability to formulate so many different areas of physics in terms of partial differential equations of the same form, Hermann von Helmholtz brings a critical attitude to bear. He points out that there is sometimes a difference in the behavior of two situations that are described by the same partial differential equations—including the same boundary conditions. The two situations to which he draws attention are: "the interior of an incompressible fluid that is not subject to friction and whose particles have no motion of rotation" and "stationary currents of electricity or heat in conductors of uniform conductivity" (Helmholtz 1891a, 58). These two configurations share the same formulation in analysis, i.e., "precisely the same" partial differential equations, and they have the same boundary conditions. Yet, their behaviors differ. Helmholtz considers, and dismisses as implausible, the explanation that the difference is a matter of the hydrodynamical equations being an "imperfect approximation to reality" (Sterrett 2017, 392). Rather, he says,

---

[12] Pincock (2012). Also all the people who have mentioned the harmonic oscillator and its several instantiations have done so, of course.

apparent contradictions between the hydrodynamic equations and "observed reality" disappear once it is recognized that discontinuous motions can occur in fluids. This is not a case of the hydrodynamic equations being wrong, though. As I put it in explaining Helmholtz's view in previous work: "The problem with the hydrodynamic equations is not that are wrong, for they are not; they are 'the exact expressions of the laws controlling the motions of fluids.' The problem is that 'it is only for a relatively few and specially simple experimental cases that we are able to deduce from these differential equations the corresponding integrals appropriate to the conditions of the given special cases.' So, the hydrodynamic equations are impeccable; it's their solution that is the problem" (Sterrett 2017, 392–3; citations from Helmholtz 1891a).

In case this seems puzzling, recall how the solution and equations they are a solution to are related. The hydrodynamic equations are governing hydrodynamic equations, but when it comes to the solution—and here he is talking about an exact solution—the solution may be expressed in terms of an equation that involves an integral. The function that is the exact solution is a function satisfying that integral equation. And evaluating that integral is where attentiveness to discontinuities in the fluid is called for, as it involves considering the range over which the pressure at every point varies. Helmholtz next considers a suggestion to simplify the problem. But he rejects that, too, as in some cases "the nature of the problem is such that the internal friction [viscosity] and the formation of surfaces of discontinuity cannot be neglected" (Helmholtz 1891b, 67). So you don't want to deal with the problem by simplifying it in a way that writes that complexity of the picture.

Another way to understand Helmholtz's point here is to consider that people often used these analogies to find solutions to equations in the way we discussed Prandtl doing above using the soap bubble membrane. For instance, someone might use an electrical circuit or setup to find the solution in a fluid flow setup. Then, Helmholtz's point would be that even though the governing partial differential equations are the same, and the boundary conditions are the same, discontinuities in fluid flow can arise in the fluid setup that will not arise in the electrical setup. The numerical value of the pressure in the fluid flow setup turns negative in the interior of the fluid, and the flow separates. Today in a practical context people would say that there is cavitation in the flow, or that the flow cavitates, as flow discontinuities tend to form.

As I explained in earlier work (Sterrett 2017, 391), the surfaces of discontinuity Helmholtz identified are an obstacle to finding a solution, too. For, as Helmholtz writes, "The discontinuous surfaces are extremely variable, since they possess a sort of unstable equilibrium, and with every disturbance in the whirl they strive to unroll themselves; this circumstance makes their theoretical treatment very difficult." Theory being of very little use in prediction here, he says, "we are thrown almost entirely back upon experimental trials, . . . as to the result of new modifications of our hydraulic machines, aqueducts, or propelling apparatus" (Helmholtz 1891b, 67). Well, what sort of experimental trials can he mean, if not the kind of analogy that he has just explained cannot be relied upon?

Helmholtz says there is another method, which he describes as follows: "In this state of affairs [the insolubility of the hydrodynamic equations for many cases of interest] I desire to call attention to an application of the hydro-dynamic equations that allows one to transfer the results of observations made upon any fluid and with an apparatus of given dimensions and velocity over to a geometrically similar mass of another fluid and to apparatus of other magnitudes and to other velocities of motion" (Helmholtz 1891b, 68). Gabriel Stokes had already, in 1850, spoken of 'similar systems' and identified conditions under which one could make inferences about similar motions and about the relation of forces in similar systems; these conditions have to do with relations between quantities in the systems (Stokes 1850, Sect. 5). In later reviews of similarity in hydrodynamics, Helmholtz's and Stokes' methods are identified as the same method, so Helmholtz is likely drawing on this earlier 1850 work of Gabriel Stokes, the same Stokes for whom the Navier-Stokes equations are named.

The method Helmholtz means here is not a matter of deduction from theory, or even of finding a solution to equations. In this paper of 1873, which would soon become foundational in empirical methods in flight research, we see that theory is still involved in the kind of inference he describes. However, the way that theory is involved is to allow someone to "transfer the results of observations made on one thing (system, machine, mass of fluid, apparatus) over to another thing (system, machine, mass of fluid, apparatus)" (Sterrett 2017, 68). It is implied, I think, that the reason this is helpful in making predictions is that some observations are more accessible, and some things are easier to manipulate and take measurements on, than others. This is reminiscent of the approach used in the realm of applied mathematics when using conformal mapping to obtain exact solutions to partial differential equations, i.e., to first transform a problem to a domain where it becomes more tractable, solve the problem in the new domain, and then transfer the solution back to the original problem. Now Helmholtz is talking about doing this with concrete, physical things, but not based on the fact that both are instantiations of the same differential equation, which, he has just shown, is not sufficient to allow one to transfer results from one setup to another. Thus, while there is an apparent similarity, Helmholtz's reasoning does not have exactly the same basis as Prandtl's soap-film method.

Helmholtz shows how one can use the governing hydrodynamic equation to which one does not have a solution to construct a mapping between two different fluids that may have different characteristics. The constraint that both of them must satisfy the hydrodynamic equations is used to determine how the various geometrical and non-geometrical quantities (time, fluid density, pressure, and coefficient of friction or viscosity) must be related. This induces a mapping (via a change of variables, as in conformal mapping) between the two fluid masses. He also distinguishes compressible from incompressible, and cohesive (liquids) from noncohesive (gases), and so on, to determine all the constants used in the change of variables that induces the mapping. In that paper, he shows how one can compare "a mass of water in which a ship is situated" and "a mass of air in which an air balloon is situated" (Helmholtz 1891b, 73). This process does use the governing hydrodynamic equation to guide the construction of the mapping, but it also formalizes the "peculiarities of air and water" in doing so, too. His approach attends to how quantities are related to each other.

In the 1873 paper, Helmholtz identifies a number of dimensionless ratios, each of which is to this day considered fundamental in establishing similarity in meterology and fluid dynamics. Dimensionless ratios are not constants.[13] Dimensionless ratios can take on various numerical values, and because the values of these ratios are informative about the thing they describe, they are often called dimensionless parameters. (A very simple case is the Mach number, which is the ratio of two velocities, hence dimensionless. Everyone is familiar with the Mach number being used to indicate whether flight is subsonic or supersonic, for instance.) Thus dimensionless parameters are informative about the similarity of two things with respect to that physical feature, and they are used to judge whether two things are similar and the ways in which they are similar. Helmholtz does not elaborate much on how one is to determine exactly what is being compared; here he uses the terms "mass of water" and "mass of air." A more general formulation of Stokes' earlier paper and Helmholtz' insight here was enabled in the early years of the twentieth century, as the field of thermodynamics developed and the notion of a system in thermodynamics (conceived of as subsuming mechanics within it) was developed.

Osborne Reynolds, Ludwig Prandtl, and Rayleigh each individually made important contributions regarding similarity in hydrodynamics worthy of in-depth discussion in their own right, and I have discussed them in a longer historical paper on the subject (Sterrett 2017, 394–397). In this chapter, we skip over them to get to the definitive statement of physically similar systems, which came from a thermodynamicist who was working as a physicist at the National Bureau of Standards: Edgar Buckingham. Though an American, Buckingham had travelled to Germany for his PhD work, working with Wilhelm Ostwald in Leipzig on a dissertation on thermodynamics. He modestly described his contribution as merely attempting to state the methods in use by researchers who used similarity methods, and to identify a rigorous basis for them, but his statement in terms of "physically similar systems" and "dimensional equations" was distinctively different from their works, and is considered the landmark work today.[14]

Buckingham took a more formal approach, one that was rooted in the nature of scientific equations: the requirement of dimensional homogeneity. It is really about the logic of equations. He was not the first to do so: Joseph Bertrand had likewise located the foundations of similarity for both mechanics and hydrodynamics in the principle of the homogeneity of equations, and attributed the insight to Isaac Newton (Bertrand 1878; Bertrand 1847). Newton had written about dimensions and units and their relation to similarity of systems, even using the term "similar systems."

[13] I mention this because I have found that, inexplicably, many philosophers think they are.

[14] Philosophers may be familiar with Buckingham's work on dimensional analysis via Percy Williams Bridgman's book *Dimensional Analysis* (Bridgman 1922). Few if any have noticed Bridgman's note in the Preface to that book expressing his indebtedness to the papers of Buckingham and to M.D. Hersey at the Bureau of Standards for presenting Buckingham's results in a series of lectures. In my entry "Dimensions" in the *Routledge Companion to the Philosophy of Physics* (Sterrett 2021), I compare their treatments, how Bridgman's treatment follows along the lines of Buckingham's, yet note what has been lost in Bridgman's partial understanding of Buckingham's very deep and philosophical work on the logic of dimensions.

In the now-landmark 1914 paper "On Physically Similar Systems: Illustrations of the Use of Dimensional Equations", Buckingham's starting point is "the most general form of a physical equation." What he means by a physical equation is an equation that describes "a relation which subsists among a number of physical quantities of n different kinds." Quantities, not variables. Dimensions, as the term is used in dimensional analysis, was developed in the context of foundational investigations into relations between quantities (e.g., Newton in investigating mechanics; Fourier in investigating heat, 19th century physicists on the relations of quantities in electromagnetism).

To get at the logic of the form of an equation that expresses a relation between different kinds of quantities, Buckingham then pares down the number of quantities by consolidating quantities of the same kind: "If several quantities of any one kind are involved in the relation, let them be specified by the value of any one and the ratios of the others to this one" (Buckingham 1914a; 345). Then, to start off simply, he restricts the discussion to cases where those ratios do not change over the course of time being considered. We are left with an equation expressing the relation between n different kinds of quantities of the form $F(Q_1, Q_2, ...Q_3) = 0$, where F is an undefined function of quantities. Further reasoning about equations in physics leads to the conclusion that every 'complete' equation of physics can be expressed in the form:[15]

$$\sum M Q_1 b_1 \cdot Q_2 b_2 ... Q_n b_n = 0$$

This is where the logic of equations of physics comes in, as this is where a principle concerning constraints on the equations of physics, i.e., the principle of dimensional homogeneity, comes in. I first give this intuitive sense of the principle: in an equation of physics, only commensurable quantities may be equated; only commensurable quantities may be added. Buckingham refers to it as "a familiar principle", credits Fourier with first stating it, and states it in his paper as follows: "all the terms of a physical equation must have the same dimensions" or, alternatively, "every correct physical equation must be dimensionally homogeneous" (Buckingham 1914a, 346).

Some ratios, such as LT^{-1} (length divided by time), will have a dimension, whereas others, such as Mach number, which is the ratio of the speed of a projectile to celerity (the speed of sound in the medium in which it is traveling) will not, since the dimension is $LT^{-1}L^{-1}T$. By the time Buckingham was writing, there were already well-known dimensionless ratios such as Mach number. These are parameters, not constants. They can take on many values, and the value they take on is often very informative (e.g., as Mach number varies from less than 1, to 1, to larger than 1, it indicates a change from subsonic to critical point to supersonic flight). Reynolds number (density · velocity · length, divided by dynamic viscosity) is likewise dimensionless and informative. In this case, it is indicative of flow regime as it proceeds from laminar to turbulent flow. In his 1914 paper, Buckingham goes on to show that, from his starting point of the most general form of a physical equation,

---

[15] See Buckingham (1914a), 346 for his explanation of what the exponents in this equation indicate.

he can derive the fact that every physical equation can be expressed in terms of dimensionless parameters, i.e., in the form $\psi(\pi_1, \pi_2, \cdots \pi_n) = 0$, where ψ is an unknown function and the dimensionless parameters π_n are independent of each other. I take the latter to mean that none of the dimensionless parameters π_n in the equation can be expressed in terms of the others (Buckingham 1914a, 347).

In a brief work reporting on his progress on the topic of physically similar systems for the first time (May 1914), Buckingham deduces the following, presenting it as a theorem about scientific equations:

> The theorem may be stated as follows: If a relation subsists among a number of physical quantities, and if we form all the possible independent dimensionless products of powers of those quantities, any equation which describes the relation is reducible to the statement that some unknown function of these dimensionless products, taken as independent arguments, must vanish. (Buckingham 1914b, 336)

I've provided the expository discussion above to help make some sense of what he says here, but for the purposes of this paper I also wish to emphasize that it is a theorem about the equations of physics, where physics is taken in a very inclusive sense. Dimensions can loosely be thought of as kinds of quantities for our purposes here.[16]

In later correspondence (to Rayleigh), Buckingham explains the role of logic and algebra as compared to the role of physical theory in his account of physically similar systems:

> I had therefore . . . to write an elementary textbook on the subject for my own information. My object has been to reduce the method to a mere algebraic routine of general applicability, making it clear that Physics came in only at the start in deciding what variables should be considered, and that the rest was a necessary consequence of the physical knowledge used at the beginning; thus distinguishing sharply between what was assumed, either hypothetically or from observation, and what was mere logic and therefore certain.[17]

Now, being able to express a physical equation as an *undetermined* function of dimensionless parameters is extremely empowering in terms of establishing similarity. In this discussion, I am interested in concrete physical models, but the use of similarity is not restricted to concrete physical models. The concept of physically similar systems can be applied to anything in physics that can be characterized as a system in the sense the term is used in thermodynamics (which includes all of classical mechanics, for classical mechanics is thermodynamics without consideration of the role of heat).

The methodology of physically similar systems enables one to use a physical model to investigate phenomena in another system, but not, as in Prandtl's use of analogy, by insight into the form of the equation describing the system behavior— and that's what is so remarkable about the method of physically similar systems. The process proceeds as follows: First, it is established that the model and what it models are physically similar systems (with respect to some relation). Usually a system S'

---

[16] I provide a more rigorous discussion in "Dimensions" (Sterrett 2021).

[17] Edgar Buckingham: Letter to Lord Rayleigh (John William Strutt) dated November 13, 1915, handwritten on official National Bureau of Standards stationery.

is constructed in such a way as to be similar to the system S, and is regarded as an experimental model of it, whether S exists in the actual world or is only a design on paper. This is done by constructing the model and setting conditions so that the values taken on by the dimensionless parameters (e.g., Mach number, Reynolds number) are the same in the model as in what it is modeling. Buckingham's discussion, while somewhat formal, provides the basis for this: "Let S be a physical system, and let a relation subsist among a number of quantities which pertain to S. Let us imagine S to be transformed into another system S' so that S' 'corresponds' to S as regards the essential quantities." He eventually deduces the nature of the similarity transformation, spelling out how one would go about setting values so that the values of the πs are the same in S as in S' (Buckingham 1914b, 353ff).

The point is elegant, reminiscent of the elegance of conformal theory: the constraint that must be satisfied in constructing the system S' is just that the value of the dimensionless parameters that appear in the general form of the equation—the arguments of the function ϕ—are the same in S' as in S. Thus, the approach Buckingham takes in constructing similar systems, the foundational basis for the construction of physically similar systems, is not a method peculiar to any particular part of physics. In that paper, he goes on to discuss applications of the method to electromagnetism (energy density of an electromagnetic field, radiation from a moving electron, and others), dynamics, and heat convection, and argues that the method is quite general. This can be very puzzling, for it doesn't seem that there is enough information in the antecedent of the theorem to permit the conclusion. Is there something about scientific equations that contributes to the argument? The answer to this is yes, and it is a matter of metrology, the science of measurement.

The requirement of a coherent system of measurement, i.e., one in which the relations between the units are the same as the relations between quantities, was adopted in the nineteenth century, and by the time Buckingham was thinking through the basis for similarity while working at the National Bureau of Standards, he could take coherence of the system of units being used in physics for granted. The logic of the quantities on which measurement systems are based is actually logically prior to the measurement system, so there is a lot of physics built into the system of measurement. The requirement that the system of units used in physics be coherent (in the above sense of the term) thus allows logical consequences to be drawn that could not be otherwise be drawn.[18]

I think it clear that, despite the spare elegance of Buckingham's account of physically similar systems, there is more at work than mathematics in accounting for the power of physically similar systems. The method he described for model experiments (experimental physical models) was based on a formalism that is in some sense even more fundamental than the mathematical equations describing the behavior of interest, and yet in some sense dependent on them: dimensional analysis, which we can think of as a formalism or language for quantities and the relations between them. Metrology and systems of measurement were developed in tandem with new developments in physics, and the use of the kinds of equations used in modern physics

---

[18] I discuss this in more detail in Sterrett (2019).

(as opposed to the proportional equations of previous eras) created the need for them. They were developed in order to enable the use of numerical interpretations of equations of physics.[19] That knowing the answer to "What are the relevant quantities involved?" is enough is striking, and is often met with incredulity. What was remarkable about Prandtl's soap-film method was that the solution to an equation could be obtained experimentally using analogy between equations. But the method of physically similar systems goes one step farther, in that one need not even have the equation describing the phenomenon of interest in hand. It is understandable that the claim is met with incredulity, unless and until the role of the coherence of a system of units for physics is appreciated.

I have two comments regarding the point that one can construct experimental physical models to investigate a phenomenon even when one does not have in hand an equation describing the phenomenon of interest.

First, the point is limited to physics (as opposed to areas of biology or sociology where one cannot draw on the same features of a system of measurement). In physics, knowing which of the quantities are relevant to a phenomenon of interest—and which are not—is actually quite a good deal of information. This is of course due to the role of the coherence of the system of units used in physics.

Second, the formulation that Buckingham provided is really very special and, I think, contingently available to us. The point about being able to do without the equation describing the phenomenon to be modeled might not have been recognized were it not for his imprint on the method.[20] The proofs and practices behind model experiments were actually not entirely new in 1914, a point he freely offered himself. But, the approach in terms of an investigation into the "most general form of physical equations"? That was new. The attention to the nature and role of the equations of physics—that attentiveness came from a physicist who had been in a community of philosophically engaged physicists who were actively discussing what units (e.g., temperature, charge) were needed in physics and what the logic of numerical scientific equations was. He had been in the thick of discussions about the role of equations in physics while studying for his doctorate in Germany, when Ostwald and Boltzmann were in dialogue. The question of whether equations were indispensable in physics, or whether, alternatively, models and analogies might do that work for the emerging physics of the day, was seriously debated (e.g., by Ludwig Boltzmann). And, then, years later, he found himself assigned the question of whether there was a proper methodology for the interpretation of model experiments, this time in the milieu of the National Bureau of Standards in Washington DC, where scientific research into establishing standards for units was being done. That he begins that investigation

---

[19] As an historical-philosophical account of equations in physics and the concomitant development of metrology, I recommend De Courtenay (2015).

[20] Note that Helmholtz, writing much earlier and in an era that predated coherence of a system of units for physics in the sense it is used here, was able to show how to establish the similarity of two physical situations, too, but that he did so by using the hydrodynamical equations. He derived dimensionless forms of the equations, and then established that if the dimensionless coefficients were the same between two situations, they would have similar motions.

with the topic of "the most general form of physical equations" is something I find noteworthy as a philosopher.

It would be a mistake to dismiss Buckingham's work, as so many philosophers have, as about "little scale models" or about engineering technology.[21] It is some of the deepest thinking about the logic of the equations of physics there is. Yes, of course, it was possible only due to all the work on similarity by other physicists, and there is no doubt that he was fortuitously located in a position unique to those writing about the basis for model experiments (experimental physical models). What should be recognized is how much more enlightened we are—or at least, could be—about the nature and role of equations, as a result of this philosophically informed account of the basis for model experiments (experimental physical models). The method of physically similar systems is not restricted to scale models, either, but is generally applicable.

2.5   Conclusion

In none of the uses of mathematics surveyed in this paper—exact solutions, simulations (numerical approximations and agent-based) and experimental physical models—is the solution to an equation simply a matter of deduction. Even in the purest example of mathematics surveyed here, i.e., exact solution of partial differential equations, the role of insight was crucial. Conceiving of the kind of mapping that might work for the problem at hand is a far cry from a straightforward application of deductive methods. This was no less true for simulations. In addition, with numerical simulations, we saw that experiential information was inextricably knit into the process by which computer simulations are produced.

We also encountered practices in science in which results that previously were thought to require an equation describing the phenomenon of interest were obtained without use of the equation. Though not news, it should give us pause that agent-based models (consisting of many agents with very simple rules) are being used to investigate behavior previously investigated using differential equations describing the behavior of continua. Most numerical methods employ a differential equation or equations describing the target behavior in some way, but agent-based models are completely different in this regard. The use of physically similar systems is another scientific practice wherein results previously thought to require having an equation describing the phenomenon of interest were obtained without the equation. Nineteenth century methods for using concrete physical models based on insights about analogies between equations were developed by Prandtl, Stokes, Helmholtz, and many others. But their similarity methods, while somewhat reliant on insight about mathematical analogies, still centered on the differential equations governing the phenomenon or behavior to be investigated. The method of physically similar

---

[21] One prominent philosopher of science, referring to Buckingham, chastised me for writing about the work of "an obscure engineer" in my book *Wittgenstein Flies A Kite* (Sterrett 2005).

systems does not. In fact, it does not require having the equation in hand in order to construct and use model experiments.

Surprisingly, it was in what one might have thought the application most dependent on practical insights, i.e., using concrete physical models, that we came across something closest to a general method for finding a solution. While it is certainly true that experiential knowledge is involved in various ways in using the method of physically similar systems, some of them nontransparently so, it is notable that the mappings (between what is to be modeled and the model, and then from model results back to what is to be modeled) can be obtained without having the equation in hand. A set of relevant dimensionless parameters can be obtained from partial knowledge, i.e., from less knowledge than is found in the equations describing the phenomenon being investigated.

It was in examining the basis for physically similar systems that an account explaining why experimental physical models can be so informative about what they model was provided. In fact, Buckingham discussed equations of physics using a whole other formalism: the language of quantity, e.g., of dimensional analysis. He used the formalism of a different kind of equation, dimensional equations, in tandem with the kinds of equations used in physics. The explanation of why the method of physically similar systems worked as well as it did, when it did, had to do with something not contained in the practices of deriving solutions either computationally or via mathematical proofs. It relied on the coherence of the system of units used in physics, which is not a matter of mathematics or logic, but is constructed apart from it, and involves both empirical results and community decisions (Sterrett 2019; De Courtenay 2015; De Courtenay 2021). It is a vast understatement to say that this point is not appreciated in philosophy of science or philosophy of mathematics. It has not gone totally unrecognized, but the work on it is seldom taken up in discussions where it would shed much light.[22]

Though this short paper is an investigation into the role of mathematics in science, it began and ended discussing equations. It has ended by recognizing a much more complex account of the equations of physics than occurred at the start (when considering exact solutions to differential equations). Because of the work on the role of dimensional equations (Buckingham 1914a, b) in showing how transformations that could solve questions about the behavior of physical systems could be answered in spite of not having an equation for that behavior, and the role of metrology (De Courtenay 2015) in enabling the use of the kinds of equations now used in physics, we can now see what we might not have realized otherwise about equations: that there is much more to them than what they say.

---

[22] De Courtenay (2015) provides an excellent account that appreciates that the role of metrology in enabling the use of numerical equations in science is well-hidden (as intended).

References

Bazant MZ (2004) Conformal mapping of some non-harmonic functions in transport theory. Proc Roy Soc Lond A 460:1433–1452

Bazant MZ (2016) Exact solutions and physical analogies for unidirectional flows. Phys Rev Fluids 1:024001

Bertrand J (1847) On the Relative Proportions of Machinery, considered with regard to their powers of working. Newton's Lond J Arts Sci 31

Bertrand J (1878) Sur l'homogénéité dans les formules de physique. Comptes rendus 86:916–920

Bridgman PW (1922) Dimensional analysis. Yale University Press, New Haven

Buckingham E (1914) On physically similar systems: illustrations of the use of dimensional equations. Phys Rev 4(4):345–376

Buckingham E (1914) The interpretation of experiments on models. J Wash Acad Sci 93:336

Castilla-Rho JC (2015) Flowlogo: an agent-based platform for simulating complex human-aquifer interactions in managed groundwater systems

De Courtenay N (2015) The double interpretation of the equations of physics and the quest for common meanings. In: Schlaudt O, Huber L (eds) Standardization in measurement, page. Pickering and Chatto, London, pp 53–68

De Courtenay N (2021) On the philosophical significance of the reform of the international system of units (SI): a double-adjustment account of scientific enquiry. Perspect Sci 1–118

Dennett D (1991) Real Patterns. J Philos 88(1):27–51

Eckert M (2019) Ludwig Prandtl: a life for fluid mechanics and aeronautical research. Springer, Dordrecht

Fillion N (2017) The vindication of computer simulation. In: Lenhard J, Carrier M (eds) Mathematics as a tool: tracing new roles of mathematics in the sciences. Boston studies in the philosophy and history of science, Springer Nature, Cham, Switzerland, pp 137–156

Harkness J, Morley F (1898) Introduction to the theory of analytic functions. Macmillan and Co, London, UK

Lenhard J (2019) Calculated surprises: a philosophy of computer simulation. Oxford University Press, Oxford

McLarty C (2023) Fluid mechanics for philosophers, or which solutions do you want for Navier-Stokes? In: Patton L, Curiel E (eds) Working toward solutions in fluid dynamics and astrophysics: what the equations don't say. Springer

Patton L (2023) Fishbones, wheels, eyes, and butterflies: heuristic structural reasoning in the search for solutions to the Navier-Stokes equations. In: Patton L, Erik Curiel (eds) Working toward solutions in fluid dynamics and astrophysics: what the equations don't say. Springer

Pincock C (2012) Mathematics and scientific representation. Oxford University Press, Oxford

Roush S (2018) The epistemic superiority of experiment to simulation. Synthese 195(11):4883–4906

Sterrett SG (2005) Wittgenstein flies a kite: a story of models of wings and models of the world. Pi Press

Sterrett SG (2017) Physically similar systems: a history of the concept. In: Magnani L, Bertolotti T (eds) Springer handbook of model-based science. Springer, Cham

Sterrett SG (2019) Relations between units and relations between quantities. In: de Courtenay N, Schlaudt O, Darrigol O (eds) The reform of the international system of units. SI. Routledge, New York

Sterrett SG (2021) Dimensions. In: Knox E, Wilson A (eds) Routledge companion to the philosophy of physics. Routledge, London

Stokes G (1850) On the effect of the internal friction of fluids on the motion of pendulums. Transactions of the Cambridge Philosophical Society, IX

Ullrich P (2005) Bernhard Riemann, thesis on the theory of functions of a complex variable (1851). In: Grattan-Guinness I, Cooke R, Corry L, Crépel P, Guicciardini N (eds) Landmark writings in western mathematics 1640–1940. Elsevier Science, Dordrecht, pp 448–459

von Helmholtz H (1868/1891) On discontinuous motions in liquids. In: mechanics of the earth's atmosphere: a collection of translations by Cleveland Abbe. Smithson. Misc Collect 843:58-66

von Helmholtz H (1873/1891) On a theorem relative to movements that are geometrically similar in fluid bodies, together with an application to the problem of steering balloons. In: Mechanics of the earth's atmosphere: a collection of translations, by Cleveland Abbe. Smithsonian Miscellaneous Collections, vol 843. The Smithsonian Institution, Washington, D.C., pp 67–77

Wilensky U (1997) NetLogo wave machine model. Center for connected learning and computer-based modeling. Northwestern University, Evanston, IL

Wilensky U (1998) NetLogo life model. Center for connected learning and computer-based modeling. Northwestern University, Evanston, IL

Wilensky U (1999) NetLogo. http://ccl.northwestern.edu/netlogo/. Center for connected learning and computer-based modeling. Northwestern University, Evanston, IL

[faded, largely illegible reference list]

Chapter 3
Fluid Mechanics for Philosophers, or Which Solutions Do You Want for Navier-Stokes?

Colin McLarty

Abstract Of the seven $1,000,000 Clay Millennium Prize Problems in mathematics, just one would immediately appeal to Leonard Euler. That is "Existence and Smoothness of the Navier-Stokes Equation" (Fefferman 2000). Euler gave the basic equation in the 1750s. The work to this day shows Euler's intuitive, vividly physical sense of mathematics.

Keywords Differential Equations · Fluid Dynamics · Navier-Stokes · Leonard Euler · Emmy Noether · Terence Tao

Patton (2023) takes the fluid mechanics of Navier-Stokes as an example of highly successful science using cryingly incomplete mathematics. She describes the current shortage of known solutions to Navier-Stokes, or even of general knowledge about not-yet-known solutions. Her argument in philosophy of science at the same time suggests more fluid philosophy of mathematics. By fluid here, I mean nimbly responsive to on-going mathematics. Well-worn philosophic debates become more vivid when posed for concepts that mathematicians are currently creating.

Fefferman (2000) does not ask whether Navier-Stokes has solutions, simpliciter. Rather Fefferman describes several notions of solution all used in current research. The Millennium Prize problem asks whether the 3-dimensional Navier-Stokes equations have *smooth* solutions. But Leray (1934) already found *weak* solutions. Weak solutions are described symbolically in Sects. 3.3–3.5 below in relation to smooth solutions and also to numerical solutions.

Many people consider weak solutions not genuine solutions. For example a weak solution might not be defined at every point of its domain. Rather, a weak solution has a well-defined *average value* in every small region. Some people claim this makes weak solutions more physically meaningful than exact solutions. Physical measurements never give point values, they give average values over small regions (Tal 2016). This is discussed in Sect. 3.6, along with a symbolic reading of Noether's

C. McLarty (✉)
Case Western Reserve University, Cleveland, Ohio, USA
e-mail: colin.mclarty@case.edu

Conservation Theorems. Whatever their physical relevance, though, weak solutions are so productive mathematically that new variations on them are still being developed in what Terence Tao describes as a "spectrum" of extensions beyond the set theoretic notion of *function* (Tao 2008a, p. 185). The themes of mathematical beauty, calculation versus constructivism, and the ontology of functions offer philosophic opportunities described in Sect. 3.7.

3.1 How the Equation Says What it Says

The Navier-Stokes equation blended symbolic calculus and physical intuition long before calculus had a rigorous logical foundation.

3.1.1 How a Fluid Moves: the Material Derivative

Consider an idealized fluid flow filling 3-dimensional space, $\mathbb{R}^3$, for time forward from 0. This posits a time-dependent fluid velocity field $\mathbf{u}$ with three components u_1, u_2, u_3 giving the velocity of the fluid flow in each direction at each point for each non-negative time. More fully, at each point $\mathbf{x} = \langle x_1, x_2, x_3 \rangle$ in $\mathbb{R}^3$ and every time $t \geq 0$ in $\mathbb{R}$, there is a velocity vector:

$$\mathbf{u}(\mathbf{x}, t) = \langle u_1(x_1, x_2, x_3, t), u_2(x_1, x_2, x_3, t), u_3(x_1, x_2, x_3, t) \rangle.$$

The change with time of the velocity field $\mathbf{u}$, at any fixed point $\mathbf{x}$, is given by the partial derivative of the velocity vector at time,

$$\frac{\partial \mathbf{u}}{\partial t} = \left\langle \frac{\partial u_1}{\partial t}, \frac{\partial u_2}{\partial t}, \frac{\partial u_3}{\partial t} \right\rangle.$$

But that derivative compares the velocities of successive particles flowing through a fixed point $\mathbf{x}$ in space. The Newtonian force law $\mathbf{F} = m\mathbf{a}$ has to describe the changing velocity, i.e. the acceleration, of one particle moving in the flow.

For that, Navier-Stokes uses the *material derivative* of each velocity component u_i along the flow $\mathbf{u}$, written Du_i/Dt. The material derivative is expressed as

$$\frac{Du_i}{Dt} = \frac{\partial u_i}{\partial t} + u_1 \frac{\partial u_i}{\partial x_1} + u_2 \frac{\partial u_i}{\partial x_2} + u_3 \frac{\partial u_i}{\partial x_3} \quad \text{or} \quad \frac{\partial u_i}{\partial t} + \sum_{j=1}^{3} u_j \frac{\partial u_i}{\partial x_j}.$$

To put that in words, in an infinitesimal amount of time t, the change in the velocity component u_i of a particle is the total of four summands. Summand $\partial u_i/\partial t$ is the change with time of the velocity component u_i at the particle's original position. If the flow is speeding or slowing there, this is one component of the acceleration of the particle. But then in addition, for each j from 1 to 3, the summand $u_j(\partial u_i/\partial x_j)$ expresses the change of the particle's velocity component u_i due to the particle's motion u_j in direction x_j.

Picture a fairly calm stream. The water flows slowly by the banks and faster in the middle. Bends and irregularities in the stream bed move water somewhat side to side. A particle of water drifting from the bank towards the middle enters a faster downstream current and so accelerates downstream.

That image corresponds to summand $u_j(\partial u_i/\partial x_j)$ in the material derivative when u_i is the downstream speed, and x_j the cross-stream coordinate. So u_j is the cross-stream speed. The partial derivative $\partial u_i/\partial x_j$ measures how the downstream speed of the current varies with the cross-stream coordinate. The product $u_j(\partial u_i/\partial x_j)$ is how much a specific particle's cross-stream speed u_j is changing that particle's downstream speed u_i.

Because the term $u_j(\partial u_i/\partial x_j)$ multiplies two unknown factors, the Navier-Stokes equation is *non-linear*. This non-linearity produces essentially all the issues discussed by (Patton 2023) and here. Appendix 3.8 gives material derivatives and the Navier-Stokes equation in the newer vector notation. Here "newer" means from the 1840s.

3.1.2   The Millennium Prize version of Navier-Stokes

In lightly adapted notation, Fefferman (2000) gives these equations for some constant $\nu > 0$ and all indices $i \in \{1, 2, 3\}$, and time t:

$$\frac{\partial u_i}{\partial t} + \sum_{j=1}^{3} u_j \frac{\partial u_i}{\partial x_j} = \nu \cdot \sum_{j=1}^{3} \frac{\partial^2 u_i}{\partial x_j^2} - \frac{\partial p}{\partial x_i} + f_i(\mathbf{x}, t) \quad (\mathbf{x} \in \mathbb{R}^3,\ t \geq 0) \quad (3.1)$$

$$\sum_{i=1}^{3} \frac{\partial u_i}{\partial x_i} = 0 \qquad\qquad\qquad (\mathbf{x} \in \mathbb{R}^3,\ t \geq 0). \quad (3.2)$$

The left side of Eq. 3.1 is the material derivative expressing fluid acceleration. That side expresses motion but no forces. Mass is cancelled out here, so Eq. 3.1 says $\mathbf{a} = \mathbf{F}/m$. The forces are on the right hand side, as described below.

Before that, though, note Eq. 3.2 is a conservation law. If $\partial u_3/\partial x_3 < 0$ at a point $\mathbf{c}$ and time t then the flow is slowing in direction x_3 as it passes $\mathbf{c}$. The fluid is either accumulating around $\mathbf{c}$ or escaping in some other direction. Conversely, if $\partial u_3/\partial x_3 > 0$ at point $\mathbf{c}$ and time t, then in direction x_3 the fluid is dispersing from $\mathbf{c}$—perhaps being replenished from other directions. Equation 3.2 rules out net accumulation or dispersal at any point. The total flow towards any point equals the total flow away. The fluid is incompressible. Beside simplifying the physical picture, incompressibility overcomes one mathematical problem posed by non-linearity (see Sect. 3.5).

Equation 3.1 says three kinds of force act on the fluid: viscous drag, pressure, and imposed external forces. Take those in reverse order. The rightmost term $f_i(\mathbf{x}, t)$ in Eq. 3.1 indicates an external force applied to the fluid, where the force varies with place and time but does not depend on how the fluid is moving. It affects the fluid motion but undergoes no feedback from the fluid. This $f_i(\mathbf{x}, t)$ is usually taken to be data, not something to be solved for. It could be gravity, or an idealized propeller producing a local current at some point in the water in an infinite basin.

The term $-(\partial p / \partial x_i)$ says this fluid flow generates and responds to pressure. But unlike familiar vacuum systems or compressed air systems, the Millenium Prize version of Navier-Stokes has no compression. Rather than being a response to compression or rarefaction of the fluid, this pressure acts instantaneously to prevent compression or rarefaction ever occurring. This is sometimes expressed by saying the speed of sound in an incompressible fluid is infinite. It is a good approximation for many practical purposes, where the speed of fluid flow is much lower than the speed of sound in that fluid. It is a bad approximation for airflow around supersonic flight. Both the compressible and the incompressible Navier-Stokes equations are used in applications. In the incompressible version used for the Millenium Prize, the force due to pressure propagates instantaneously.

Most interesting is the viscosity term $\nu \cdot \sum_{j=1}^{3}(\partial^2 u_i / \partial x_j^2)$. This *Laplacian*, the sum of unmixed second partials of u_i, measures how the average of u_i on a small sphere around any point differs from the value at that point. See Sect. 3.1.3 below, and Appendix 3.9. This pretty mathematics is not the only physically defensible mathematical model of viscosity and not the only one used in applications. It was Navier's model, and Stokes's, among others, and it is now the standard basic model in pure and applied fluid mechanics.

The Navier-Stokes equation assumes a positive scalar ν:

$$\nu \cdot \sum_{j=1}^{3} \frac{\partial^2 u_i}{\partial x_j^2} \qquad \nu > 0.$$

So it says each particle is pulled towards the average fluid velocity around it. If ν is close to zero that means the fluid velocity near a point has very little effect on the motion at that point. It means a weak pull and a not-very viscous fluid. If ν is large this is a strong pull and very viscous fluid. The scalar ν is the inverse to the *Reynolds number* often used in applications, and in an application it can be found experimentally. This leads into the important practical, theoretical, and conceptual problems of scaling in fluid mechanics discussed by Sterrett (2006, pp. 150ff.).

The Millenium Prize problem is to prove, or disprove: For every smooth 0-divergence velocity field $\mathbf{u}(\mathbf{x}, 0)$ given as initial condition, there are smooth fields $\mathbf{u}(\mathbf{x}, t)$ and $p(\mathbf{x}, t)$ solving the equations. Fefferman (2000, p. 2) gives full details.

3.1.3   Invariance

Appendix 3.9 easily adapts to show these mathematical conditions characterize the Laplacian $\sum_{j=1}^{3}(\partial^2 u_i/\partial x_j^2)$ uniquely, up to a non-zero scalar multiple:

1. It is non-zero (i.e. viscosity does induce some forces).
2. It has 0 constant term (i.e. viscosity induces no force if the fluid is not moving).
3. It is linear in **u** and its derivatives (the non-linearity of the material derivative is bad enough).
4. It is invariant under coordinate rotation.
5. It has the lowest degree among all operators with the other properties (specifically it uses only second derivatives).

These differ in their physical plausibility. Properties 1 and 2 are obvious features of viscosity. Properties 3 and 5 are mathematically convenient, and experience shows good differential equations in physics are often linear with degree one or two.

Property 4 is the richest. Cognoscenti say viscous drag is diffusion of momentum. If the average momentum in direction u_i around point **x** is higher than it is exactly at **x**, then it diffuses towards **x**. The property says the diffusion depends only on the difference between the momentum at **x** and the average in its environs. It does not depend at all on *where* around **x** the momentum is higher, or lower. Momentum in this model diffuses with the flow just as strongly as across the flow, and against the flow. This is neither implausible nor obvious. Different 19th century mathematicians gave different justifications (Darrigol 2002).

3.2   The Question of Solutions

> Till a few decades ago (early 1930s), there was unanimous opinion that the Navier-Stokes equations were useful (in agreement with the experiments, that is) only at "low" velocity regimes. It is also thanks to the efforts of outstanding mathematicians such as Jean Leray, Eberhard Hopf, Olga Ladyzhenskaya, and Robert Finn that they are nowadays regarded as the universal foundation of fluid mechanics. (Galdi 2011, p. 3)

The change is not due to finding many exact solutions to the equations nor even (as the Millennium Prize attests) to proving many exist. Patton (2023) argues that current philosophy of science poorly reconciles the success of the Navier-Stokes equation in pure and applied fluid mechanics, with the shortage of known solutions. So Patton argues for a *heuristic* interpretation of this fluid mechanics, and other sciences using very difficult mathematics. This argument, in turn, calls for exploring the several senses of "solutions" to differential equations. The case of Navier-Stokes is compelling because so little is known of its solutions in any sense. Curiel (2019) argues the Einstein Field Equations in General Relativity are as compelling.

3.2.1   Contrast the Question of Numbers

Mathematicians rarely ask *What are numbers?* since, say, Hilbert and Brouwer in the 1920s. Not because they don't care. To the contrary mathematicians use numbers so constantly and in such sophisticated ways that every community of mathematicians has a conception of numbers suited to their needs. With the mid-20th century rise of structural mathematics, mathematicians arrived at a standard, rigorous structural concept of numbers. It just works differently than many "structuralist" philosophers of mathematics prefer (McLarty 2008, 2017). In contrast, mathematicians continue articulating ever further answers to this next question.

3.2.2   What are Solutions to PDEs?

Often, when a problem is easily solved in many ways, those ways relate easily to each other, so the differences require no critical attention. (Patton 2023) highlights how far Navier-Stokes is from that case, and how this matters in theory and applications.

Contrast an easy Partial Differential Equation (PDE), the Laplace equation:

$$\frac{\partial^2 u}{\partial x^2} + \frac{\partial^2 u}{\partial y^2} + \frac{\partial^2 u}{\partial z^2} = 0.$$

One solution is

$$u(x, y, z) = \frac{1}{(x^2 + y^2 + z^2)^{1/2}} \quad \text{when } \langle x, y, z \rangle \neq \langle 0, 0, 0 \rangle.$$

This solution is:

1. Exact (it is a well defined function on $\mathbb{R}^3$ omitting the origin $\{\mathbf{0}\}$).
2. Symbolic (expressed by explicit elementary functions and verifiable as a solution by cookbook calculus).
3. Numerically manageable (computer routines give accurate approximations, with well understood bounds on their accuracy).

The numerical manageability comes in part from the well understood exact solution. In numerical approximations it is important to know when to apply a very fine grid, to a fast changing function, so as to closely track the true value. But it is equally important to avoid too fine a grid. Too fine a grid, applied to a slow-changing function, leads to so many calculations of such tiny differences that round-off errors will overwhelm the true values. This exact solution to the Laplace equation has obvious fast and slow changing regions. But even for harder, more complicated problems, better understanding of exact solutions guides better choices in numerical work.

A stated goal of Bazant and Moffat's work on exact, albeit very artificial, solutions is to "provide useful test cases for numerical simulations" (2005, p. 55).

Let these ideas be the basis for a tentative[1] typology of solutions to PDEs:

1. *Exact solutions* are functions, sufficiently differentiable on some stated domain, satisfying the equations.
2. *Symbolic solutions* are explicit devices with specified calculating rules verifying the equations. These *could* be expressed by familiar functions with explicit calculating rules (polynomials, trig functions, exponentials…) but they *need not* actually be functions at all.
3. *Numerical solutions* are calculating procedures for approximating numerical values of solutions, with reasonable bounds on accuracy.

Approximations are meaningless without some estimate of accuracy. For simple practical applications the estimate may simply be: "it works okay." But even an utterly practical application may be too complex, expensive, irreproducible, or dangerous to just "try it and see how well it works." Estimating accuracy is often the hardest part of giving approximations.

For Navier-Stokes, in the present state of knowledge, these three desiderata for solutions pull in different directions. Some of that conflict may be intrinsic to the problem. There are many important open questions here and a correspondingly huge mathematical and engineering literature.

It should be clear at the outset that the following sort of problems could arise. In current work on Navier-Stokes they constantly do arise:

1. An exact solution may be unsymbolic—provably not expressible by any combination of standard, named functions.[2] And chaotic turbulence, for example, may make numerical approximations meaningless.
2. A symbolic solution may not represent any actual function.
3. A numerical procedure might give no theoretical or intuitive insight. Notably, if it blows up for certain initial conditions, it may be impossible to judge whether that reflects actual physical instability in those conditions or is a mere artefact of the computational procedure (Fefferman 2000, p. 2).

As of now, *weak solutions*, described below, are by far the favored approach to finding exact solutions, and central to numerical solutions. Weak versions of Navier-Stokes (or any PDE) are the first step in discretization for numerical calculation by the *finite element method*. Weak solutions could play a large role in articulating heuristic understandings of Navier-Stokes as Patton (2023) urges for philosophy of science. They are certainly key to the current mathematical ontology of PDEs.

---

[1] At the conference "What Equations Don't Say" (July 2020) I called this trichotomy "not yet successful or complete," a phrase lifted from Patton's description of some fluid mechanics. This is not offered as opposed to Sterrett's exact solutions, approximations, simulations, and models, but for development with those.

[2] Tao has posted on the importance of explicit solutions, and more, at "Why global regularity for Navier-Stokes is hard" terrytao.wordpress.com/2007/03/18/.

3.3   Weak Derivatives

A major strategy for symbolic solution of differential equations takes rules which are rigorously true for differentiable functions, and applies them *regardless of whether the functions in question have derivatives in the standard ϵ-δ sense*. This leads to calculating with what may not be functions at all. It would quickly give contradictions if done naively. So it is not done naively.

The pivotal rule here is *integration by parts*:

$$\int (f' \cdot g) = f \cdot g - \int (f \cdot g').$$

To see what it says rearrange the terms:

$$f \cdot g = \int (f \cdot g') + \int (f' \cdot g).$$

Then it is just the integral of the product rule for derivatives:

$$(f \cdot g)' = f \cdot g' + f' \cdot g.$$

All of these equations are rigorously correct when f and g have continuous derivatives. But they make no conventional sense if f or g has no derivative.

Practically, this technique is a mainstay of first year calculus, written:

$$\int_a^b f'(x)g(x)\,dx = f(x)g(x)\Big|_a^b - \int_a^b f(x)g'(x)\,dx. \tag{3.3}$$

Sometimes the easiest way to integrate a function $h(x)$ is to factor it as $f'(x)g(x)$ for some functions f, g such that the switched product $f(x)g'(x)$ is easy to integrate. Calculus textbooks give examples as do numerous websites.

More relevant to theory, integration by parts is a staple in deriving laws of mathematical physics. Darrigol (2002) shows constant appeals to it in the history of Navier-Stokes. And it is central to all avatars of calculus of variations including routine 19th century practice, the long sophisticated project to install Weierstrassian rigor, and Noether's formal version.

The point in symbolic solution of PDEs is that the right hand side of Eq. 3.3 can be well defined, in the conventional ϵ-δ sense, even when f has no derivative in that sense. Then the left hand term $f'(x)g(x)$ can serve to give a symbolic meaning to $f'(x)$, called the *weak derivative* of f. Symbolic integration by parts characterizes weak derivatives f'', $f^{(3)}$, $f^{(4)}$, ... of all finite orders even when f has no conventional derivative at all, so long as the other functions g have enough conventional derivatives.

If a function f has an actual derivative f' in the ϵ-δ sense then f' will also be a weak derivative. If f has no actual derivative at some point then the weak derivative

f' may not be an actual function around that point—while it may still be an actual function away from that point. Compare the Dirac delta $\delta(x)$ in Appendix 3.10.

A technical obstacle has been a major research focus since at least Laurent Schwartz in the winter of 1944 (Barany 2018) and up to active research today (Tao 2008a).[3] Symbolic calculations with Eq. 3.3 depend on the functions g having good conventional derivatives and integrals. Those functions are called *test functions*, and a central fact of current analysis is that no single sense of "test function" works for all current uses of weak derivatives. An often used class of test functions is:

Definition 1 A *compactly supported test function* on any coordinate space $\mathbb{R}^n$ is any real-valued function $g(x)$ which is smooth (has derivatives of all finite orders) and compactly supported (meaning $g(x) = 0$ for all x outside some finite region of $\mathbb{R}^n$).

Two things are important for us here:

1. A compactly supported function $g(x)$ has constant value 0 everywhere outside a finite region. So we can reasonably say even for symbolic functions $f(x)$, the product $f(x)g(x)$ is 0 outside that region. Thus we can reasonably say

$$f(x)g(x)\Big|_{-\infty}^{\infty} = 0 - 0 = 0.$$

2. If $g(x)$ is a compactly supported test function then its derivative $g'(x)$ is too.

Then Eq. 3.3 says for all compactly supported test functions g, and all actual or symbolic functions $f(x)$:

$$\int_{-\infty}^{\infty} f'(x)g(x)\,dx = -\int_{-\infty}^{\infty} f(x)g'(x)\,dx. \tag{3.4}$$

This is called "(formally) integrating f' against g by parts."

These considerations using test functions have motivated various precise ideas of *generalized functions*. This section and Appendix 3.10 will give some basics on generalized functions but, for reasons discussed in Sects. 3.3.1 and 3.4.1, we do not define generalized functions in terms of compactly supported test functions. We use an intuitive somewhat indefinite notion of *generalized function* $f(x)$. We stipulate only that they include a lot of actual functions (smooth or not), all have weak derivatives as characterized by Eq. 3.4, and integration of all generalized functions $f_1(x)$, $f_2(x)$ retains the chief good properties familiar from the Riemann integral of actual, continuous functions:

- $\int_a^b f_1(x)\,dx + \int_a^b f_2(x)\,dx = \int_a^b (f_1(x) + f_2(x))\,dx.$
- $\int_a^b k \cdot f_1(x)\,dx = k \cdot \int_a^b f_1(x)\,dx$ for all $k \in \mathbb{R}.$
- $\int_a^b f_1(x)\,dx + \int_b^c f_1(x) = \int_a^c f_1(x)\,dx$
  when $a < b < c \in \mathbb{R} \cup \{-\infty, \infty\}.$

---

[3] Research with Schwartz on theoretical aspects of this was Alexander Grothendieck's dissertation (1952). Grothendieck's functorial methods here presaged his later ideas in cohomology.

This symbolic core is common to all theories of generalized functions from Fourier in the 1820s to today. Appendix 3.10 uses this core to derive some general properties of weak derivatives, and to show the Dirac delta function $\delta(x)$ is the weak derivative of the Heaviside step function $\theta(x)$.

3.3.1 The Problem of Multiplication

There is a problem with multiplying weak derivatives by one another, or even by actual functions g that are not smooth. Equation 3.4 does not cover such multiplication. This is today a grave problem for work on non-linear PDEs. But weak derivatives are so valuable in the non-linear case that this is a research problem, briefly sketched in Sect. 3.4.1, and not a reason to abandon the method.

3.4 Weak Solutions to Differential Equations

Deane Yang (Courant Institute) well says:

> The study of nonlinear PDEs is almost always done in an ad hoc way. This is in sharp contrast to how research is done in almost every other area of modern mathematics. Although there are commonly used techniques, you usually have to customize them for each specific PDE. (MathOverflow July 24, 2020)

A differential equation on an unknown function $f(x)$ can use derivatives of any order $f'(x)$, $f''(x)$ .... And there may be functions $f(x)$ which almost solve it except that some derivative $f^{(n)}$ in the equation is undefined at some points. Experience since the 18th century shows it can be useful to just proceed anyway as if the derivatives of $f(x)$ are not a problem.

Today the results of that idea use weak derivatives, when standard ϵ-δ derivatives are not known to exist, and are called *weak solutions*. They became a major tool in solving PDEs long before they were systematized. In particular, Jean Leray (1934) showed the Navier-Stokes equation has weak solutions for all smooth initial conditions (and more) at a still-early stage of the idea. There is no fully general definition of weak solutions in the non-linear case today.

Equation 3.4 offers two approaches to weak solutions:

Introduction strategy   Use weak derivatives $f'(x)$, $f''(x)$, ... with the symbolic calculational meaning given by Eq. 3.4.

Elimination strategy   Use symbolic integration against an arbitrary compactly supported test function $g(x)$, following Eq. 3.4, to eliminate any order derivative $f^{(n)}(x)$ in the equation in favor of $\int f(x)g^{(n)}(x)\,dx$.

Section 3.5 sketches the usual elimination strategy for Navier-Stokes, already used by Leray. In contrast, that strategy is unavailable for the Einstein Field Equations in General Relativity.

3.4.1 Distributions

> The theory of distributions provides a most satisfactory framework to generalized solutions
> in linear theory. The question of what is a good concept of a generalized solution in nonlinear
> equations, though fundamental, is far more murky. (Klainerman 2010)

There are many variant ways to define weak or generalized solutions. All are complicated in detail. All these approaches have the simple goal of making Eq. 3.4 describe weak derivatives for generalized functions.

Further definition is needed to say what generalized functions there are. The basic one today uses compactly supported test functions. We will call the corresponding generalized functions *distributions*. For full definitions see numerous reference works in print and on line, and for history Barany (2018).

These distributions work wonderfully for linear equations. Notably, if f is a weak solution in this sense to a linear differential equation, and f is in fact a smooth function, then f is an exact solution to the differential equation. This very desirable property fails in general for non-linear equations. But the weak form of Navier-Stokes in Sect. 3.5 has this property. Numerous variant senses of weak solution are sketched by Tao (2008a, b). Perhaps the most important approach today uses *Sobolev spaces*, an important idea we cannot explore here.

3.5 The Weak Navier-Stokes Equation

> To arrive at the idea of a weak solution of a PDE, one integrates the equation against a
> test function, and then integrates by parts (formally) to make the derivatives fall on the test
> function. (Fefferman 2000, p. 3)

The resulting integral equation is called a *weak form* of the original PDE. Solutions to the weak form are *weak solutions* to the original PDE. The procedure works straightforwardly for linear PDEs. As a sample consider the linear term $\partial u_i / \partial t$ at the start of Eq. 3.1. Integrating against a test function v gives the left hand term here, and then formal partial integration turns it to the right hand term:

$$\iint_{\mathbb{R}^3 \times \mathbb{R}} \frac{\partial u_i}{\partial t} \cdot v \, dx \, dt \quad = \quad - \iint_{\mathbb{R}^3 \times \mathbb{R}} u_i \cdot \frac{\partial v}{\partial t} \, dx \, dt.$$

That presents no problems.

However the non-linear term $u_j(\partial u_i / \partial x_j)$ poses a problem. Integrating this against a test function v gives

$$\iint_{\mathbb{R}^3 \times \mathbb{R}} u_j \cdot \frac{\partial u_i}{\partial x_j} \cdot v \, dx \, dt.$$

Partial integration does not authorize taking the derivative $\partial/\partial x_j$ off of u_i and putting it on v while simply ignoring the factor u_j. That would be incorrect even if $\mathbf{u}$ was an actual smooth vector field.

As quoted from Yang in Sect. 3.3, the general procedure must be somehow customized for Navier-Stokes. Equation 3.2 supplies a fine way which is most easily expressed in the vector form, introduced in Appendix 3.8. In Eq. 3.7 the non-linear term appears as $\mathbf{u} \cdot \nabla \mathbf{u}$. But Eq. 3.8 says $\nabla \cdot \mathbf{u} = 0$, which implies

$$\mathbf{u} \cdot \nabla \mathbf{u} = \nabla \cdot (\mathbf{u} \otimes \mathbf{u}).$$

So, instead of integrating $\mathbf{u} \cdot \nabla \mathbf{u}$ against the vector test function $\mathbf{v}$, we can integrate $\nabla \cdot (\mathbf{u} \otimes \mathbf{u})$ against it. That gives sums of terms as on the left here. These do allow partial integration:

$$\iint_{\mathbb{R}^3 \times \mathbb{R}} \frac{\partial (u_i u_j)}{\partial x_j} v_i \, dx \, dt = -\iint_{\mathbb{R}^3 \times \mathbb{R}} u_i u_j \frac{\partial v_i}{\partial x_j} \, dx \, dt.$$

This weakens Navier-Stokes in a very nice way, such that all the steps are provably correct for smooth vector fields $\mathbf{u}$ and p.

In full, Fefferman takes any *compactly supported smooth vector field* $\mathbf{v}$ and writes a vector integral equation:

$$-\iint_{\mathbb{R}^3 \times \mathbb{R}} \mathbf{u} \cdot \frac{\partial \mathbf{v}}{\partial t} \, dx \, dt - \sum_{ij} \iint_{\mathbb{R}^3 \times \mathbb{R}} u_i u_j \frac{\partial v_i}{\partial x_j} \, dx \, dt = \qquad (3.5)$$

$$\nu \iint_{\mathbb{R}^3 \times \mathbb{R}} \mathbf{u} \cdot \nabla^2 \mathbf{v} \, dx \, dt + \iint_{\mathbb{R}^3 \times \mathbb{R}} \mathbf{f} \cdot \mathbf{v} \, dx \, dt + \iint_{\mathbb{R}^3 \times \mathbb{R}} p \cdot (\nabla \cdot \mathbf{v}) \, dx \, dt$$

A vector field $\mathbf{u}$ and scalar field p satisfying this equation for all test functions $\mathbf{v}$, and satisfying the analogues for Eq. 3.2, is a weak solution to Navier-Stokes. A key fact here is that any weak solution $\mathbf{u}$, p where these are actual smooth functions, is an exact solution. This rests on the particular relation between Eq. 3.1 and 3.2. No such nice weakening is available for the Einstein field equations, let alone for general non-linear PDEs.

Equation 3.5 is more complicated than Eq. 3.1 but more suited to systematic analysis. Leray (1934) showed all divergence free initial conditions (which may be smooth or may themselves be weak) have weak solutions in this sense.

However, people study weak solutions $\mathbf{u}$, and their weak derivatives $\partial u_i/\partial x_j$, by deeper analogues of the calculations in Appendix 3.10. One goal is to know where $\mathbf{u}$ is smooth or at least continuous. Many efforts have achieved many kinds of success at this, as Lemarié-Rieusset (2015) describes. A key problem for Navier-Stokes (and for nonlinear PDEs in general) remains that non-linearity means the u_i and $\partial u_i/\partial x_j$ are multiplied together—which cannot be done with distributions.

Some yet-unknown ad hoc adjustments will be needed before this approach to exact solutions of Navier-Stokes will work very generally. Or, maybe some other approach will work first.

3.6   Relations to Physics and Engineering

> The landmark idea in the theory of weak solutions is to give up on solving the equations point-wise but trying to solve them in an averaged sense, which is meaningful also from a physical point of view. In the case of fluid mechanics, we expect a very complex behavior by (turbulent) flows appearing in real life, hence we expect to be able to capture only averages of the velocity and pressure. (Berselli and Spirito 2021, p. 1)

Imagine a wind tunnel experiment measuring, for example, the air flow around a model of an airplane wing. The measuring instruments never find a velocity component $u_i(x_1, x_2, x_3, t)$ at a single geometric point $\mathbf{x}$ and point of time t. Rather, the instruments register some kind of net effect (we might loosely say "average") of the flow in direction x_i in a little region around point $\mathbf{x}$ for a little time around t.

Mathematically, "net effect" is expressed as an integral. The net effect of u_i nearby to $\mathbf{x}, t$ is isolated mathematically by integrating u_i against a *bump function* g:

$$\iint_{\mathbb{R}^3 \times \mathbb{R}} u_i(x_1, x_2, x_3, t) \cdot g(x_1, x_2, x_3, t) \, dx \, dt \qquad (3.6)$$

The bump function g is a special kind of compactly supported test function. It has constant value 1 on some little region U around $\mathbf{x}, t$, and dies off quickly (but smoothly) to 0 within some slightly larger region V containing U, and then is constant 0 outside V. So the product $u_i \cdot g$ equals u_i in region U, and is 0 far from U. In the small gap between U and V, $u_i \cdot g$ lies between u_i and 0.

Of course, those conditions allow many choices of g. Just as a wind tunnel experiment will use some specific selection and placement of measuring devices, so one mathematical analysis of a function might use a specific selection of bump functions. The general theory of weak solutions uses not only bump functions but all compactly supported test functions g (or some substitute for those depending on exactly what kind of weak solutions are wanted). Since weak solutions rest on this kind of integral, some people regard them as solutions "in an averaged sense."

Then there is, *prima facie*, a disadvantage of generalized solutions in physics, namely non-uniqueness. A physical law normally has a unique solution for given initial conditions, telling what will follow after those initial conditions. (Even in quantum mechanics, the time evolution of the wave function is determinate, outside of collapse.) Current methods of generalized solutions to PDEs typically give far from unique results. We can go no farther into this than to say "this unsatisfactory situation may be temporary due to our technical inabilities or unavoidable in the sense that the concept itself is flawed" (Klainerman 2010).

As a second, possible, physical advantage: failure of smoothness in solutions to a PDE may not be a defect. Black holes in General Relativity show singularities may just be a fact. The Clay Institute will also award the $1,000,000 prize for a proof that Navier-Stokes has no smooth solution for some smooth 0-divergence initial conditions. If Navier-Stokes has no smooth global solution for some smooth initial condition, then any Leray weak global solution for that condition conceals singularities which could be revealed.[4] That might be informative.

This may be unimportant, though, if Navier-Stokes in fact has smooth solutions for all smooth initial conditions. (Klainerman 2010) finds an "almost universal assumption that these equations are globally regular." But (Tao, 2016) finds reasons they might not be. The solutions in Bazant and Moffatt (2005) have singularities, but they begin with singularities in their initial conditions.

3.6.1   Inexact Solutions and Mathematical Freedom

Weak solutions give precise form to a standard 19th century working style which remains common enough today. In engineering contexts it is often associated with Oliver Heaviside who lived 1850 to 1925. See the sly fusion of unobtrusive modern rigor with swift formal calculation using shifted delta functions $\delta(x - T)$ and step functions $H(x)$ and so on in the innovative textbook Strang (2015).

Sterrett (2006, p. 106) quotes a vivid statement by Ludwig Boltzmann (born 1844 died 1906) arguing about the use of models in science: "all partial differential equations of mathematical physics [are] only inexact schematic pictures for definite areas of fact."

Weierstrass would not agree. Knowing the physics might be inexact, he would insist the equations per se are precise statements. But Boltzmann knew most of his colleagues would find his imagery obviously true. Even in the purest of mathematics, most would freely manipulate such "pictures" by familiar rules, blithely ignoring the ϵ-δ definitions of analysis. If some particular case led to obvious nonsense, then they would step back and be more careful with that case. Rough and ready methods often gave valuable results. Today weak solutions combine some of that freedom with all the rigor Weierstrass could have wanted.

3.6.2   Formal Calculus of Variations

Emmy Noether (b. 1882–d. 1835) described her conservation theorems as theorems of *formale Variationsrechnung*, which is to say *formal calculus of variations* (1918).

---

[4] The converse does not hold. Leray's weak solutions are now known to be not unique (Albritton et al. 2022). Some Leray weak solution for a given initial condition might have singularities while some other for those same initial conditions does not.

$$\sum \psi_i \, \delta u_i \; = \; \delta L \; -$$

$$-\frac{d}{dx}\left\{ \sum \left(\binom{1}{1} \frac{\partial L}{\partial u_i^{(1)}} \, \delta u_i + \binom{2}{1} \frac{\partial L}{\partial u_i^{(2)}} \, \delta u_i^{(1)} + \cdots + \binom{\kappa}{1} \frac{\partial L}{\partial u_i^{(\kappa)}} \, \delta u_i^{(\kappa-1)} \right) \right\}$$

$$+\frac{d^2}{dx^2}\left\{ \sum \left(\binom{2}{2} \frac{\partial L}{\partial u_i^{(2)}} \, \delta u_i + \binom{3}{2} \frac{\partial L}{\partial u_i^{(3)}} \, \delta u_i^{(1)} + \cdots + \binom{\kappa}{2} \frac{\partial L}{\partial u_i^{(\kappa)}} \, \delta u_i^{(\kappa-2)} \right) \right\}$$

$$\cdots + (-1)^\kappa \frac{d^\kappa}{dx^\kappa}\left\{ \sum \binom{\kappa}{\kappa} \frac{\partial L}{\partial u_i^{(\kappa)}} \, \delta u_i \right\}.$$

Fig. 3.1 Noether's Lagrangian equation

Despite considerable scholarship no one is entirely sure what she meant by that (Brading and Brown 2010; Kosman-Schwarzbach 2011; McLarty 2018). Very likely she meant it as a symbolic calculus.

These famous theorems have since been made exact by constructions on tangent and cotangent bundles developed after 1950. These can fairly be taken as formalizing her ideas (Olver 1986). But very likely she herself construed her basic equation symbolically. She wrote it for one independent variable x as in Fig. 3.1.

Noether's chief motivation was in General Relativity with four independent variables for spacetime. Yet even she (or possibly her printer?) balked at printing the much longer equation for several independent variables.

The "first variations" in this equation, written δL and δu_i, had formal calculating rules at the time, as did another variation operator Δ used in Noether's proof. Probably Noether in 1918 just followed those which she would learn from Kneser (1904). Her terribly swift arguments are inexplicit by the standards of her time and standards today (Kosman-Schwarzbach 2011; Noether 1918).

Numbers 19 and 23 of the famous Hilbert Problems aimed at installing Weierstrassian rigor in the calculus of variations (Hilbert 1902). Several of Noether's colleagues at Göttingen worked on this complicated challenge. Noether never thought her theorems needed it.

3.7 Philosophical Opportunities

The general principles upon which Euler bases himself are the Newtonian laws expressed in differential form, the complete acceptance of the concept of force, the use of pressure as force per surface unit, and the use of clearly defined systems of Cartesian coordinates. All are expressed with absolute conceptual clarity. (Calero 2008, p. 401)

3.7.1 Beautiful Mathematics

This is beautiful mathematics. The poor rudiments given here barely scratch the surface of 265 years of clear mathematical conceptions and elegant formulations. I do not know how to prove it is beautiful but I will cite the fact that Darrigol stopped counting after he found five major re-discoveries of the viscosity term. The earlier parts of the equation were given by Euler in 1755 (Darrigol 2002, p. 98). Experts immediately accepted Euler's equation as a Newtonian account of incompressible, inviscid fluid flow. But it made a poor match to observed fluid motions. Viscosity became more and more clearly the chief problem so a viscosity term had to be discovered. The Navier-Stokes one was discovered, over and over, first by Navier in 1822, and among others by Stokes in 1845. Darrigol (2002, p. 50) reports "as late as 1860" the Navier-Stokes equation "could still be rediscovered." Researchers often specifically rejected each other's physical justifications, and all considered earlier attempts inadequate. None succeeded very far in terms of practical hydraulics. Yet they repeatedly arrived at this mathematics.

To this day, despite all the well known difficulties of using Navier-Stokes, and the very many variants and alternatives that coexist in practical applications because of those difficulties, people do not merely continue working with Navier-Stokes. Navier-Stokes was chosen by leading mathematicians to be the oldest one of just seven problems in all of mathematics to qualify for a $1,000,000 Millenium prize.

This is enduring mathematical beauty. Some will prefer to avoid the idea of beauty and say this viscosity term was discovered repeatedly because it has the very nice invariance properties described in Sect. 3.1.3 above. Instead of dismissing beauty, though, I suggest this low complexity, high symmetry, and extreme difficulty explains the beauty of Navier-Stokes. These properties do explain the equation's long-lasting influence and its centrality to current fluid mechanics.

3.7.1.1 Visual Beauty

As another aspect of beauty, Patton titled her conference talk "Fishbones, Wheels, Eyes, and Butterflies: A Heuristic Account of Models and Simulations" evoking the marvelous color graphics of solutions to Navier-Stokes in Bazant and Moffatt (2005).[5] Similar color photographs of actual gas jets are produced by *Planar Laser-Induced Fluorescence* (PLIF) and can be easily found online. There has been good philosophic work on graphics and diagrams in mathematics (De Toffoli 2017). There is less on aesthetics in the sense of eye-appeal. Is the vivid eye appeal of color especially helpful for the challenge of seeing 3-dimensional fluid motion? This kind of color graphic is also used to depict the rate of motion of complex numbers away from Julia sets in complex dynamics (Gleick 1987).

---

[5] Talk given at the Midwest Philosophy of Mathematics Workshop 17, University of Notre Dame, 12–13 November 2016.

3.7.2   Responsible Fictionalism

As a last point of esthetics, the beauty of the equations clearly deviates from physical truth. While incompressibility is a reasonable idealization for many purposes, it is strictly incompatible with the existence of sound waves or shock waves in the fluid. Also, no physical fluid has quite constant viscosity under changing pressure. This can be practically important in industry. These idealizations often need adjustment before they work in practice. As to theory, Darrigol shows at length how difficult the issue of fluid continuity was for the founders since all believed actual matter is molecular.

This suggests a question for philosophical fictionalists who say mathematical objects are not real but are useful fictions (Bueno 2009; Leng 2010). Can and should people more assertively claim authorship of these fictions? Is Euler not only the author of beautiful equations but also of continuous, incompressible fluids?

3.7.3   Calculation

Philosophers should better notice the difference between symbolic and numerical calculation. Both get too little attention from philosophers of mathematics—compared to logical foundations, and high level concepts.[6] My own work focuses on foundations and high-level concepts. Wonderful topics. But they are wonderful not in contrast to calculation. They are wonderfully related to calculation.

The key example of calculation for us here is Bazant and Moffatt (2005) on Navier-Stokes and the result it is based on, (Bazant 2004, p. 1436). The striking innovation was seeing *what* to prove. The *proofs* are explicit vector-calculus calculations with analytic functions. They are finally artful use of rules listed in tables at the end of calculus texts, such as:

$$\frac{\partial\,fg}{\partial x} = \frac{\partial f}{\partial x}g + f\frac{\partial g}{\partial x} \qquad \frac{d}{dt}\,f(t, x(t), y(t)) = \frac{\partial f}{\partial t} + \frac{\partial f}{\partial x}\frac{dx}{dt} + \frac{\partial f}{\partial y}\frac{dy}{dt}.$$

The rules give symbolic algorithms—as the name *calculus* was originally intended. They can be, and often are, applied by laptop computers using environments like Mathematica® or MATLAB®. Numerous free websites under names like "integral calculator" or "differential equation calculator" will solve problems using these rules. These free sites devote very limited resources to the work and of course they can fail to solve a problem even when it is solvable. On the other hand, Appendix 3.9

---

[6] (Fefferman 2000, p. 4) concludes "Standard methods from PDE appear inadequate to settle [Navier-Stokes]. Instead, we probably need some deep, new ideas." He does not mean abstract or logically high-order ideas like forcing, or derived functor cohomology. Fefferman and Tao won Fields Medals for deep insights based on novel, logically concrete concepts (Carlson 1980; Fefferman 2007).

gives a calculation on the Laplacian operator that those websites can do by symbolic routine, but humans can do quickly and insightfully by a geometric trick.

Bazant and Moffatt's calculations are exact, rigorous, application of standard rules to well defined expressions. They are as explicit, and exact as

$$\sin^2 \omega + \cos^2 \omega = 1 \quad \text{or} \quad e^{x+y} = e^x \cdot e^y.$$

They are neither approximate nor numerical. They are exact and symbolic–though in turn they lend themselves to the numerical approximations used in generating the graphics for (Bazant and Moffatt 2005).

Sophisticated computer proof programs have gotten good philosophic attention (Avigad 2020). But the routine, formulaic, and really symbolic calculations done on paper, or laptop computers, have been too little appreciated.[7]

3.7.3.1   Distinguishing Constructivism From Calculation

Philosophers routinely fail to see the gap between *constructive methods* producing solutions "calculable in principle," and actual calculations. Actual calculations will use any means available to get done. Both historically and today, pursuing the "calculable in principle" is very different from calculating.

Leopold Kronecker, who lived 1823 to 1891, was a constructivist before the term existed. He was clear that he would accept only problem solutions showing explicitly how to calculate specific answers—in principle. But he did not produce many calculations in fact. His key algorithms were infeasible beyond trivial cases before digital computers. In contrast, Paul Gordan (living 1837 to 1912) is often called a constructivist but he was not. He neither had nor sought any idea of "calculations in principle" versus non-constructive existence proofs. He sought and produced vast, actual calculations. He cared about, and wrote about, what he could and could not calculate in fact (McLarty 2012).

Mathematicians who make hard actual calculations are rarely constructivists. They use general results based on excluded middle, the Hahn-Banach theorem, whatever, whenever those generalities help to justify a hard calculation.

A constructive exact solution to the Navier-Stokes equation can be far from concrete. It may just not be expressible by familiar, explicit functions. And, while it provides means of calculating numerical approximations in principle, it may give absolutely no feasible means.

---

[7] This connects to current logic research. The calculations discussed here could probably be well formalized at logical strength too low to even register in current Reverse Mathematics (Simpson 2010). They are done on laptop computers (i.e. finite state machines).

3.7.4 Ontology of Mathematics

In analysis, it is helpful to think of the various notions of a function as forming a spectrum, with very "smooth" classes of functions at one end and very "rough" ones at the other. (Tao 2008a, p. 185)

I have spoken of functions in the set theoretic sense, such as the sine function $\sin(x)$ and the exponential e^x, and spoken of other function-like things which are not functions set-theoretically, like the Dirac delta $\delta(x)$ and other weak derivatives f' of functions f which have no derivatives in the ϵ-δ sense.

Tao urges that in current analysis it is best to think of all these, and much more, as kinds of functions. His spectrum of notions of function includes functions in the usual set-theoretic sense. The infinitely differentiable ones are classically called "smooth," and Tao calls them smooth. Those smooth functions that furthermore are limits of their Taylor series are classically called "analytic" and Tao calls them "very smooth." Set-theoretically well-defined functions which lack ϵ-δ derivatives or even are discontinuous, Tao calls rougher. The Dirac delta and other distributions, and several more constructs which are set-theoretically not functions at all, Tao calls rougher yet. In a nutshell, these rougher "functions" do not have well-defined values at points (and may have provably no possible value at some points) but have average values over intervals. Tao (2008a) compares several currently important ways of doing this, and leads up to major open questions.

Tao is a major architect of current research in analysis. But some of the "generalized" functions used today already existed in analysis before anyone gave the now standard set-theoretic definition of function. There is a well developed historical literature. Lützen (1982) shows how Joseph Fourier in 1822 had the idea of the Dirac delta function, which Jean-Gaston Darboux later rejected as not a function. There is a huge advanced literature today in theoretical mathematics, numerical methods, physics, and engineering. To emphasize the centrality of generalized functions to current, undergraduate math see how *step* and *impulse* solutions organize the entire Calc IV textbook Strang (2015) right from the start (p. 21).

This in itself could draw the attention of philosophers of mathematics. What is more, Tao's blog https://terrytao.wordpress.com gives an expert real-time view of developing ideas. This together with distinctive undergraduate exposition like (Tao 2014) offers a rare degree of access to live mathematics in the making.

To close, though, I note traditional ontological questions in philosophy of math become concrete and vivid in relation to Patton's *heuristic* interpretation of Navier-Stokes theory. Beyond debates over whether the number 2 exists—these questions can illuminate what mathematicians actually do:

1. Are weak derivatives like $\delta(x)$ objects, or just calculating terms? (Re Appendix 3.8, the same question for the vector ∇.)
2. Are numerical solutions really solutions, or merely instructions? That is, are they objects or just calculating procedures?
3. Does mathematics use some altogether more fluid ontology than *objects*, *terms*, and *procedures*?

3.8   Appendix: Symbolic Vector Calculus

This article uses cartesian coordinates $\langle x_1, x_2, x_3 \rangle$ and partial derivatives $\partial/\partial x_i$. Much of the literature today uses vector notation with the *del* or *nabla* operator ∇ created by William Rowan Hamilton in the 1840s.[8] Darrigol (2002, p. 97) explains the historical and conceptual issues, and his choice to use ∇. The body of this essay uses partial derivatives for brevity. The chief advantage of ∇ is in calculating and we do little of that. Fefferman (2000) elegantly combines the notations.

Equation 3.1 is actually three equations, one for each of u_1, u_2, and u_3. In vector notation, that and 3.2 become truly two equations (Bazant and Moffatt 2005, p. 55). The left hand side of Eq. 3.7 is the material derivative $D\mathbf{u}/Dt$ in vector form:

$$\frac{\partial \mathbf{u}}{\partial t} + \mathbf{u} \cdot \nabla \mathbf{u} = \nu \cdot \nabla^2 \mathbf{u} - \nabla p + \mathbf{f}, \tag{3.7}$$

$$\nabla \cdot \mathbf{u} = 0. \tag{3.8}$$

The ∇ operator is a symbolic vector of partial derivative operators.

$$\nabla = \langle \partial/\partial x_1, \partial/\partial x_2, \partial/\partial x_3 \rangle.$$

Given function p (for pressure) the *gradient* ∇p is the symbolic multiple of vector ∇ by scalar p. Of course $\partial p/\partial x_1$ is really not the multiple of $\partial/\partial x_1$ by p:

$$\nabla p = \langle \partial p/\partial x_1, \partial p/\partial x_2, \partial p/\partial x_3 \rangle.$$

The *divergence* of a vector field $\mathbf{u}$ is often written div($\mathbf{u}$), or $\nabla \cdot \mathbf{u}$ as in Eq. 3.8. It is a symbolic dot product of vectors:

$$\nabla \cdot \mathbf{u} = \langle \partial/\partial x_1, \partial/\partial x_2, \partial/\partial x_3 \rangle \cdot \langle u_1, u_2, u_3 \rangle = \frac{\partial u_1}{\partial x_1} + \frac{\partial u_2}{\partial x_2} + \frac{\partial u_3}{\partial x_3}.$$

For a vector field $\mathbf{u}$, the *gradient* $\nabla \mathbf{u}$ is the matrix whose columns are the gradients of the components of $\mathbf{u}$. It is the symbolic *outer product* $\nabla \otimes \mathbf{u}$:

$$\nabla \mathbf{u} = \nabla \otimes \mathbf{u} = \begin{bmatrix} \partial u_1/\partial x_1 & \partial u_2/\partial x_1 & \partial u_3/\partial x_1 \\ \partial u_1/\partial x_2 & \partial u_2/\partial x_2 & \partial u_3/\partial x_2 \\ \partial u_1/\partial x_3 & \partial u_2/\partial x_3 & \partial u_3/\partial x_3 \end{bmatrix}.$$

This is the transpose of the *Jacobian* matrix of $\mathbf{u}$. The term $\mathbf{u} \cdot \nabla \mathbf{u}$ in Eq. 3.7 means the product of $\mathbf{u}$ as a row vector with $\nabla \mathbf{u}$ as a 3×3 matrix.

Section 3.5 uses the *divergence* of a 3×3 matrix $\mathbf{A} = [A_{ij}]$ in the form of the row vector of divergences of the column vectors of $\mathbf{A}$. So it is the symbolic matrix

---

[8] See Miller (1997) and references there.

product $\nabla \cdot \mathbf{A}$. That section invokes a result that can be checked by calculation: For any vector field $\mathbf{u}$,

$$\nabla \cdot (\mathbf{u} \otimes \mathbf{u}) = (\nabla \cdot \mathbf{u}) \cdot \mathbf{u} + \mathbf{u} \cdot \nabla \mathbf{u}.$$

3.9 Appendix: The Laplacian

The *Laplacian* showcases the unity of geometry and formalism in classical calculus. In three variables it is the differential operator

$$\frac{\partial^2}{\partial x_1^2} + \frac{\partial^2}{\partial x_2^2} + \frac{\partial^2}{\partial x_3^2}.$$

It is often written ∇^2 because it is a symbolic dot product $\nabla \cdot \nabla$:

$$\langle \partial/\partial x_1, \partial/\partial x_2, \partial/\partial x_3 \rangle \cdot \langle \partial/\partial x_1, \partial/\partial x_2, \partial/\partial x_3 \rangle = \sum_{j=1}^{3} \frac{\partial^2}{\partial x_j^2}.$$

The Laplacian of a function f at a point $\langle a, b, c \rangle$:

$$\left[\frac{\partial^2 f}{\partial x_1^2} + \frac{\partial^2 f}{\partial x_2^2} + \frac{\partial^2 f}{\partial x_3^2} \right]_{\langle a,b,c \rangle}$$

measures how the average of f on a small sphere around $\langle a, b, c \rangle$ differs from the value $f(a, b, c)$ at that point.

To see this, shift the coordinates if necessary so the point $\langle a, b, c \rangle$ is $\langle 0, 0, 0 \rangle$. And shift the value of f by a constant amount, if necessary, so that $f(0, 0, 0) = 0$. The average value of f on the sphere S_r around $\langle 0, 0, 0 \rangle$, with radius r, is the integral of f over S_r divided by the sphere area. We will see for small enough r this average is $r^2/6$ times the Laplacian at 0:[9]

$$\frac{1}{4\pi r^2} \iint_{S_r} f \, dS = \frac{r^2}{6} \left[\frac{\partial^2 f}{\partial x_1^2} + \frac{\partial^2 f}{\partial x_2^2} + \frac{\partial^2 f}{\partial x_3^2} \right]_{(0,0,0)}. \tag{3.9}$$

To shorten the notation write x, y, z rather than x_1, x_2, x_3 and write

$$f_x = \frac{\partial f}{\partial x} \quad f_{xx} = \frac{\partial^2 f}{\partial x^2} \quad f_{xy} = \frac{\partial^2 f}{\partial x \partial y} \quad \text{and so on.}$$

---

[9] The reader can adapt this to a modern proof of equality in the limit as r goes to 0, when f is integrable on some neighborhood of 0 and is twice differentiable at 0.

Take r small enough that terms of order 3 (and higher) in the Taylor series of f at 0 are negligible on S_r. Inside this sphere $f(x, y, z)$ equals

$$f_x(0) \cdot x + f_y(0) \cdot y + f_z(0) \cdot z + \frac{f_{xx}(0) \cdot x^2}{2} + f_{xy}(0) \cdot xy + \qquad (3.10)$$

$$\frac{f_{yy}(0) \cdot y^2}{2} + f_{yz}(0) \cdot yz + \frac{f_{zz}(0) \cdot z^2}{2} + f_{xz}(0) \cdot xz.$$

Here is the key: Every term in Formula 3.10 except those for x^2, y^2, and z^2 has integral 0 over S_r. The values of $f_x(0) \cdot x$ on the hemisphere of S_r with $x \geq 0$ are exactly the negatives of those on the hemisphere with $x \leq 0$, so over the whole sphere it integrates to 0. The same hemispheres show $f_{xy}(0) \cdot xy$ and $f_{xz}(0) \cdot xz$ both integrate to 0. Either variable y or z works for $f_{yz}(0) \cdot yz$. Only the squares x^2, y^2, and z^2 remain. So the integral of f over S_r is

$$\frac{1}{2} \iint_{S_r} \left[f_{xx}(0) \cdot x^2 + f_{yy}(0) \cdot y^2 + f_{zz}(0) \cdot z^2 \right] dS.$$

Routine calculation, or a geometric shortcut, shows for constants A, B, C

$$\iint_{S_r} \left[Ax^2 + By^2 + Cz^2 \right] dS = \frac{4}{3}\pi r^4 (A + B + C).$$

Applying this to the constants $f_{xx}(0)$, $f_{yy}(0)$, $f_{zz}(0)$ proves Eq. 3.9.

The shortcut: By spherical symmetry the integral of x^2 over S_r equals that of y^2 or z^2. So it is one third of the integral of $x^2 + y^2 + z^2$. But on this sphere $x^2 + y^2 + z^2$ is the constant r^2. So the integral of x^2 over S_r is $\frac{1}{3}r^2$ times the area of S_r, thus $\frac{4}{3}\pi r^4$.

3.10 Appendix: Some Weak Derivative Calculations

Some facts about weak derivatives follow directly from the symbolic core of the idea given in Sect. 3.3. For example that core implies

$$(f_1(x) + f_2(x))' = f_1'(x) + f_2'(x).$$

Weak derivatives need not be actual functions, so this equation cannot be proved by evaluating the two sides. It is proved by showing the two sides have the same formal integral against every test function g. See Fig. 3.2. A careful expositor might say this proves $(f_1 + f_2)'$ and $f_1' + f_2'$ are *weakly equal*, or *equal in the sense of distributions*. But insofar as the derivatives only exist symbolically, or as distributions, there is no other way they could be equal.

$$\int_{-\infty}^{\infty} (f_1(x)+f_2(x))' \cdot g(x)\, dx \;=\; -\int_{-\infty}^{\infty} (f_1(x)+f_2(x)) \cdot g'(x)\, dx$$

$$= \; -\int_{-\infty}^{\infty} f_1(x) \cdot g'(x)\, dx - \int_{-\infty}^{\infty} f_2(x) \cdot g'(x)\, dx$$

$$= \; \int_{-\infty}^{\infty} f_1'(x) \cdot g(x)\, dx + \int_{-\infty}^{\infty} f_2'(x) \cdot g(x)\, dx$$

$$= \; \int_{-\infty}^{\infty} (f_1'(x) + f_2'(x)) \cdot g(x)\, dx.$$

Fig. 3.2 Verifying $(f_1 + f_2)' = f_1' + f_2'$

The well known Dirac delta $\delta(x)$ is defined by saying, for all $g(x)$,

$$\int_{-\infty}^{\infty} \delta(x) \cdot g(x)\, dx \;=\; g(0). \tag{3.11}$$

It is widely popularized as a function with $\delta(x) = 0$ when $x \neq 0$, and $\delta(0)$ so high that the area under the graph is 1. This is not meant literally.[10] No finite or infinite value for $\delta(0)$ makes definition 3.11 work by any standard definition of integral. The Dirac delta $\delta(x)$ is not an actual function.

In fact, $\delta(x)$ is useful largely because it is the weak derivative of an actual function. That is often called Heaviside's step function $\theta(x)$, defined by:

$$\theta(x) = \begin{cases} 0 & \text{if } x < 0 \\ 1 & \text{if } 0 \leq x. \end{cases}$$

Again, $\theta(x)$ is an actual function with an actual value at every point x. But it is discontinuous at 0 and so has no derivative at 0 in the standard ϵ-δ sense.

Yet the weak derivative $\theta'(x)$ is proved equal to $\delta(x)$ by the symbolic integration in Fig. 3.3. The first line is an instance of Eq. 3.4. Then come standard calculations with the actual functions θ, g, and g'; then the definition of $\delta(x)$. Thus $\theta'(x) = \delta(x)$, at least weakly or in the sense of distributions. There is no other way they could be equal since neither one is an actual function.

3.10.1 Going Farther Into $\delta(x)$

While $\delta(0)$ has no assignable value, there is a clear sense to saying $\delta(x) = 0$ for all $x \neq 0$. Take any open interval $(a, b) \subset \mathbb{R}$ with $0 \notin (a, b)$, and any test function $g(x)$

---

[10] It can be taken literally, with some care, in non-standard analysis (Katz and Tall 2013).

$$\int_{-\infty}^{\infty} \theta'(x) \cdot g(x)\, dx \; = \; -\int_{-\infty}^{\infty} \theta(x) \cdot g'(x)\, dx$$

$$= \; -\int_{-\infty}^{0} \theta(x) \cdot g'(x)\, dx - \int_{0}^{\infty} \theta(x) \cdot g'(x)\, dx$$

$$= \; 0 - \int_{0}^{\infty} g'(x)\, dx \; = \; -g(x)\Big|_{0}^{\infty} \; = \; g(0) \; = \; \int_{-\infty}^{\infty} \delta(x) \cdot g(x)\, dx.$$

Fig. 3.3 Verifying $\theta' = \delta$

such that $g(x) = 0$ for all x outside of (a, b). Then also $\delta(x) \cdot g(x) = 0$ outside that interval, and so:

$$\int_{a}^{b} \delta(x) \cdot g(x)\, dx \; = \; \int_{-\infty}^{\infty} \delta(x) \cdot g(x)\, dx \; = \; 0.$$

The *fundamental lemma of calculus of variations* says constant 0 is the only *continuous* function on (a, b) giving integral 0 with all test functions g. Thus constant 0 is the unique continuous function representing $\delta(x)$ away from $x = 0$.

Analogous use of well chosen test functions reveals information about weak solutions in more advanced cases. (Fefferman 2000, pp. 3–4) surveys the results up to that time on weak solutions of Navier-Stokes.

Recommended Reading

Two concise classics make vector calculus vivid: *Div, Grad, Curl, and All That* by Schey (1992) oriented toward applications, and *Calculus on Manifolds* by Spivak (1971) toward theory. For generalized functions see "impulse" and related topics throughout Strang (2015) oriented toward applications, and Tao (2008a) toward theory. Lemarié-Rieusset (2015) discusses Navier-Stokes largely in the Millennium Prize version. Jeremy Gray pointed out a conceptual survey every philosopher pursuing the math or physics of PDEs should read: (Klainerman 2010).

Recommended Viewing

The movie "Gifted" (2017) shows a mother, daughter, and granddaughter each facing Navier-Stokes. It flashes back to the first two in changing conditions for women in recent mathematics, with Hollywoodesque optimism for the future. There is actual math in a funny bit about the Gaussian integral.

References

Albritton D, Bruè E, Colombo M (2022) Non-uniqueness of Leray solutions of the forced Navier-Stokes equations. To appear in the Ann Math. Abstract online at annals.math.princeton.edu/articles/18745

Avigad J (2020) Modularity in mathematics. Rev Symb Log 13(1):47–79

Barany M (2018) Integration by parts: wordplay, abuses of language, and modern mathematical theory on the move. Hist Stud Nat Sci 48:259–299

Bazant M (2004) Conformal mapping of some non-harmonic functions in transport theory. Proc R Soc Lond A 460:1433–54

Bazant M, Moffatt K (2005) Exact solutions of the Navier-Stokes equations having steady vortex structures. J Fluid Mech 541:55–64

Berselli L, Spirito S (2021) On the existence of Leray-Hopf weak solutions to the Navier-Stokes equations. Web J Fluids. https://doi.org/10.3390/fluids6010042

Brading K, Brown H (2010) Symmetries and Noether's theorems. In: Brading K, Castellani E (eds) Symmetries in Physics, Philosophical Reflections, chapter 5. Cambridge University Press, pp 89–109

Bueno O (2009) Mathematical fictionalism. In: Bueno O, Linnebo Ø (eds) New waves in philosophy of mathematics. Palgrave Macmillan

Calero JS (2008) The genesis of fluid mechanics 1640–1780. Studies in History and Philosophy of Science, Springer, Netherlands

Carlson L (1980) The work of Charles Fefferman. In: Lehto O (ed) Proceedings of the International Congress of Mathematicians. vol 1. Academia Scientiarum Fennica, Helsinki, (1978), pp 53–56

Curiel E (2019) On geometric objects, the non-existence of a gravitational stress-energy tensor, and the uniqueness of the Einstein Field equation. Stud Hist Philos Mod Phys 66:90–102

Darrigol O (2002) Between hydrodynamics and elasticity theory: the first five births of the Navier-Stokes equation. Arch Hist Exact Sci 56:95–150

De Toffoli S (2017)'Chasing' the diagram–the use of visualizations in algebraic reasoning. Rev Symb Log 10(1):158–86

Fefferman C (2000) Existence and smoothness of the Navier Stokes equation. Clay Mathematical Institute, Cambridge, MA

Fefferman C (2007) The work of Terence Tao. In: Sanz-Solé M, Soria J, Varona JL, Verdera J (eds) Proceedings of the International Congress of Mathematicians. vol 1. European Mathematical Society, Madrid, Spain (2006), pp 78–87

Galdi G (2011) An introduction to the mathematical theory of the Navier-Stokes equations: steady-state problems, 2nd edn. Springer, New York

Gleick J (1987) Chaos: making a new science. Penguin Books, USA

Grothendieck A (1952) Résumé des résultats essentiels dans la théorie des produits tensoriels topologiques et des espaces nucléaires. Ann l'Institut Fourier, Grenoble 4:73–112

Hilbert D (1902) Mathematical problems. Bull. Amer. Math. Soc. 8:437–479

Jacobson N (ed) (1983) E. Noether: Gesammelte Abhandlungen. Springer, New York

Katz M, Tall D (2013) A Cauchy-Dirac delta function. Found Sci 18:107–23

Klainerman S (2010) PDE as a unified subject. In: Alon N, Bourgain J, Connes A, Gromov M, Milman V (eds) Visions in mathematics. Birkhäuser, Basel, pp 279–315

Kneser A (1904) Variationsrechnung. In: Encyklopädie der mathematischen Wissenschaften mit Einschluss ihrer Anwendungen, vol 2, T.1, H.1, Teubner, pp 571–641

Kosman-Schwarzbach Y (2011) The Noether theorems: invariance and conservation laws in the twentieth century. Les Éditions de l'École Polytechnique. Springer, New York. Trans. Bertram E. Schwartzbach.

Lemarié-Rieusset P (2015) The Navier-Stokes problem in the 21st century. Taylor & Francis

Leng M (2010) Mathematics and reality. Oxford University Press, Oxford UK

Leray J (1934) Sur le mouvement d'un liquide visqueux emplissant l'espace. Acta Math 63:193–248

Lützen J (1982) Prehistory of the theory of distributions, in Studies in the History of Mathematics and the Physical Sciences, vol 7, Springer, New York

McLarty C (2008) "There is no ontology here": visual and structural geometry in arithmetic. In: Mancosu P (ed) The Philosophy of mathematical practice. Oxford University Press, pp 370–406

McLarty C (2012) Theology and Its discontents: David Hilbert's foundation myth for modern mathematics. In: Doxiadis A, Mazur B (eds) Mathematics and narrative. Princeton University Press, Princeton, pp 105–129

McLarty C (2017) The roles of set theories in mathematics. In: Landry E (ed) Categories for the working philosopher. Oxford University Press, pp 1–17

McLarty C (2018) The two mathematical careers of Emmy Noether. In: Beery J (ed) Women in mathematics: 100 years and counting. Springer, pp 231–52

Miller J (1997) Earliest uses of symbols of calculus. jeff560.tripod.com/calculus.html

Noether E (1918) Invariante Variationsprobleme. Nachrichten von der Gesellschaft der Wissenschaften zu Göttingen, pp 235–57. In Jacobson 1983, pp 248–70

Olver PJ (1986) Applications of Lie groups to differential equations. Springer, New York

Patton Lydia (2023) Fishbones, wheels, eyes, and butterflies: heuristic structural reasoning in the search for solutions to the Navier-Stokes equations. In: Patton Lydia, Curiel Erik (eds) Working toward solutions in fluid dynamics and astronomy: what the equations don't say. Springer

Schey H (1992) Div, Grad, Curl, and all that. WW Norton

Simpson S (2010) Subsystems of second order arithmetic. Cambridge University Press

Spivak M (1971) Calculus on manifolds. Westview Press

Sterrett S (2006) Wittgenstein flies a kite: a story of models of wings and models of the world. Pi Press

Strang G (2015) Differential quations and Linear Algebra. Wellesley-Cambridge Press

Tal E (2016) How does measuring generate evidence? the problem of observational grounding. J Phys: Conf Ser 772:012001

Tao T (2008a) Distributions. In: Gowers T, Barrow-Green J, Leader I (eds) Princeton companion to mathematics. Princeton University Press, pp 184–87

Tao T (2008b) Function spaces. In: Gowers T, Barrow-Green J, Leader I (eds) Princeton companion to mathematics. Princeton University Press, pp 210–13

Tao T (2014) Analysis I. Hindustan Book Agency, New Delhi

Tao T (2016) Finite time blowup for an averaged three-dimensional Navier-Stokes equation. J Am Math Soc 29:601–74

Chapter 4
Fishbones, Wheels, Eyes, and Butterflies: Heuristic Structural Reasoning in the Search for Solutions to the Navier-Stokes Equations

Lydia Patton

Abstract Arguments for the effectiveness, and even the indispensability, of mathematics in scientific explanation rely on the claim that mathematics is effective or necessary in successful scientific predictions and explanations. Well-known accounts of successful mathematical explanation in physical science appeals to scientists' ability to solve equations directly in key domains. But there are spectacular physical theories, including general relativity and fluid dynamics, in which the equations of the theory cannot be solved directly in target domains, and yet scientists and mathematicians do effective work there (McLarty 2023, Elder 2023). Building on extant accounts of structural scientific explanation (Bokulich 2011, Leng 2021), I argue that philosophical accounts of the role of equations in scientific explanation need not rely on scientists' ability to solve equations independently of their understanding of the empirical or experimental context. For instance, the process of formulating solutions to equations can involve significant appeal to information about experimental contexts (Curiel 2010) or of physically similar systems (Sterrett 2023). Working from a close analysis of work in fluid mechanics by Martin Bazant and Keith Moffatt (2005), I propose an account of heuristic structural explanation in mathematics (Einstein 1921, Pincock 2021), which explains how physical explanations can be constructed even in domains where basic equations cannot be solved directly.

Keywords Explanation · Equations · Fluid dynamics · Navier-Stokes · Simulation · Physics

L. Patton (✉)
Virginia Tech, Blacksburg, VA, USA
e-mail: critique@vt.edu

4.1   The Argument from Successful Theories

Contemporary structuralist and structural realist accounts analyze the stability of the formal relations within a theory, arguing from the persistence and success of an entity or relation ('structure') in scientific theories to the objectivity, or even the reality, of those elements. According to Hilary Putnam, if Newton simply thought up the law of gravity, and that law, with its consequences, deep embedding in other relations, and extraordinary fruitfulness sprang into existence at Newton's whim, that would be a miracle (Putnam 1975, 73). Instead, the best explanation of the success of the law is that Newton caught on to something, and the fruitfulness and depth of the relation he caught on to is evidence that his theory consists of statements about the world with "objective content" (73). Putnam went on to argue that certain formal relations and elements of mathematics are not only stable across theory change, not only fruitful and deep as a theory develops over time, but also are indispensable to the results of physical theories.[1] Resnik makes a pragmatic indispensability argument on this basis for the existence of mathematical objects and the truth of mathematical statements:

1. In stating its laws and conducting its derivations science assumes the existence of many mathematical objects and the truth of much mathematics.

2. These assumptions are indispensable to the pursuit of science; moreover, many of the important conclusions drawn from and within science could not be drawn without taking mathematical claims to be true.

3. We are justified in drawing conclusions from and within science only if we are justified in taking the mathematics used in science to be true. (Resnik 1995, 169–70)

One can turn such arguments around and apply them to the physical theories in which mathematical reasoning is applied. According to "methodological" arguments for scientific realism, those elements essential to successful scientific theories warrant realist commitment. Methodological realism about the mathematics employed in scientific theories is closely related to indispensability, and works roughly as follows:

1. Scientists work with formal posits (e.g., equations, variables and relations), and with mathematical reasoning in general, in proving results.

2. Some of these formal relations persist, and are indispensable to the success of our best scientific theories.

3. In such cases (2), commitment to the reality of the entities and relations that are indispensable to the success of our theories is justified.

The key argument is from the success of a theory to realism. To run an indispensability argument requires determining which elements of a theory are required to make the results of the theory come out true.

---

[1] Putnam's original 'no miracles' argument was made in the context of mathematics, not physics. The indispensability argument is sometimes associated with Quine as well. Colyvan (2015, Sect. 1) emphasizes that Quine and Putnam took this position to be a matter of intellectual honesty. We are committed to the existence of quarks, electrons, neutrinos, and so on. In the case of abstract entities or relations, this commitment may only be because they are employed in successful scientific theories. But, in that case, we must also be committed to the existence of the real number line.

The 'Divide et Impera' ('divide and conquer' or DEI) strategy is a version of methodological realism.[2] DEI argues from the necessary employment of a structure or posit in a theory to its objectivity or reality.[3] Structural realism is a version of the DEI strategy (Psillos 2018, Sect. 2.7).[4] Mathematical structure is necessary to derive the results of a theory, and that structure persists over theory change.

These accounts rely on a set of assertions which add up to a proposed argument, which can be called "The Leading Argument".

The Leading Argument

1. Theory change is a Darwinian process in which the most successful theories win out over the less fruitful or explanatory ones.

2. Theories are successful if they are able to generate more predictions or explanations in the target domain than rivals.[5]

3. Theories that "latch on" to objective relations or to real elements will be more successful than rivals that do not.[6]

    a. The elements that are truly necessary to prove the results of the theory should be isolated from the 'idle wheels' that do not contribute to the theory's success. (Methodological realism)

    b. The mathematical structure of a theory is likely to be deployed in successful theories, to be clarified and refined as theories develop, and to be an effective instrument for constructing scientific explanations.[7]

4. The best explanation for the persistence of structures or elements over theory change is that the entities, processes, or structures to which the theory refers are objective or real. (The no miracles argument)

5. Mathematical structure persists over theory change.

Conclusion Mathematical structure that persists over theory change is more likely to be objective or to refer to reality.

---

[2] See Kitcher 1993, Psillos 1999, Leplin 1986. See Lyons 2006, Cordero 2011, Patton 2015 for responses.

[3] "The underlying thought," as Psillos notes, "is that the empirical successes of a theory do not indiscriminately support all theoretical claims of the theory, but rather the empirical support is differentially distributed among the various claims of the theory according to the contribution they make to the generation of the successes" (Psillos 2018, Sect. 2.5).

[4] "In opposition to scientific realism, structural realism restricts the cognitive content of scientific theories to their mathematical structure together with their empirical consequences. But, in opposition to instrumentalism, structural realism suggests that the mathematical structure of a theory represents the structure of the world (real relations between things)... structural realism contends that there is continuity in theory-change, but this continuity is … at the level of mathematical structure" (Psillos 2018, Sect. 2.7).

[5] A main theme of Kitcher's *The Advancement of Science* (Kitcher 1993).

[6] Worrall (Worrall 1989) famously argues for this claim; a version of this argument can be found in Peirce and the pragmatist tradition.

[7] The indispensability or 'unreasonable effectiveness' argument (the latter due to Eugene Wigner). A classic assertion of structural realism, but found in other approaches as well.

A number of challenges have been posed to arguments from the success of science to the reality or truth of mathematics.[8] W. v. O. Quine's 'web of belief' view denies that mathematical or logical claims are more epistemically fundamental than empirical assertions based on observation. Instead, Quine argues, abstract mathematical or logical statements simply become more central to our web of belief: more statements depend on them than vice versa.[9] The centrality of mathematical and logical claims within our conceptual schemes makes us less inclined to give them up, and we might be less likely to question them than we are to question other beliefs. But that does not make mathematical or logical claims more epistemically certain or more likely to be true than claims based on observation. When a theory makes a false prediction, ultimately the only evidence about how to revise the theory comes from "our own direct observations".[10]

One might question whether the employment of abstract entities and mathematics in physical reasoning provides them with a warrant of objectivity or reality at all. Recent accounts emphasize that models can be deliberately counterfactual, or even fictional, and can still serve their desired epistemic roles. Alisa Bokulich (2017) emphasizes that idealized, even counterfactual models can support successful scientific reasoning, which may challenge the argument from the use of models in successful theories to a warrant of those models' truth or reference to reality.[11]

Both these challenges are articulated against the background assumption that successful theories or models are successful for epistemic reasons. That assumption itself can be questioned. In analyzing episodes such as the Chemical Revolution, Hasok Chang notes that claims that one theory is more epistemically successful than another may be based on epistemic grounds, or on something "cruder": "an unreflective triumphalism that celebrates the winning side in an episode, whichever it may happen to be" (Chang 2009, 240). Chang argues that, from a contemporary

---

[8] Laudan's well-known argument against convergent realism has been covered extensively, so I won't discuss it in detail here.

[9] "Mathematics and logic, central as they are to the conceptual scheme, tend to be accorded [...] immunity, in view of our conservative preference for revisions which disturb the system least; and herein, perhaps, lies the 'necessity' which the laws of mathematics and logic are felt to enjoy" (Quine 1950, xiii). Thus, while Quine is often associated with the 'indispensability' argument, he did not argue that indispensability made mathematics more likely to be necessary or epistemically certain beyond the possibility of revision.

[10] When faced with a theory that is newly in conflict with observation, "Toward settling just which beliefs to give up we consider what beliefs had mainly underlain the false prediction, and what further beliefs had underlain these, and so on [...] We stop such probing of evidence, as was remarked, when we are satisfied. Some of us are more easily satisfied than others. [...] But there is a limit: when we get down to our own direct observation, there is nowhere deeper to look. [...] the ultimate evidence that our whole system of beliefs has to answer up to consists strictly of our own direct observations" (Quine and Ullian 1978, 13).

[11] To be sure, accounts that untether models from their targets have deep questions to answer. As Knuuttila notes, "Any account approaching scientific models as fictions faces at least the two following challenges. First, if scientific models are considered as fictions rather than representations of real-world target systems, how are scientists supposed to gain knowledge by constructing and using them? And, second, how should the ontological status of fictional models be understood?" (Knuuttila 2021, 5078).

perspective, Lavoisier's theory "was just as wrong as the phlogiston theory in its advanced versions" (240). It makes little sense to argue that Lavoisier's theory was more successful because it was truer or more epistemically justified: from a modern perspective, that is not the case.

We will focus on the question: What counts as success for a scientific theory? The leading argument relies on the empirical claim that mathematics must be deployed in successful theories and explains their success. According to this argument, mathematical and logical structures are necessary to the success of physical theories in generating predictions and explanations. When we focus on the use of differential equations, a problem leaps out at once: differential equations are not directly solvable in many domains of interest.[12] Given that fact, how are we to evaluate claims that the mathematics in question is 'unreasonably effective'[13] in those domains?

4.2 Are Theories Based on Differential Equations Successful?

The leading argument relies on a history of successful scientific theories in which mathematical structures are indispensable, that is, are necessary to deriving predictions and explanations within a theory.[14] Premises 2 and 3b of the leading argument rest on a claim about the development of science: that theories are successful if they predict and explain, and that mathematical structure is more likely to be deployed in successful theories. What happens, then, if precisely the problem that faces scientists is the inability to solve the equations, the inadequacy of current theory to make predictions and support explanations in the domain of interest on the basis of solved equations?

In that case, indispensability arguments lose much of their force for these theories. If we cannot derive the desired results directly from the equations, which are central to a theory's mathematical structure, then how can we argue that structural mathematical reasoning is effective, much less necessary or indispensable? How can we compare theories to each other on the basis of which results are provable in which, which is necessary to decide which is more successful?

No problem, we might assert confidently. The leading argument runs for successful theories. But theories in which direct solutions to a theory's equations are available

---

[12] My thinking on this issue has deep roots in long conversations with Erik Curiel about this topic. To learn more about Curiel's view, I warmly recommend Curiel 2010, Curiel 2016. After giving this paper as a talk in Indiana, discussions with Colin McLarty and Susan Sterrett have been indispensable to developing these views further. Their own contributions are found in McLarty 2023, Sterrett 2023.

[13] To use Wigner's phrase.

[14] Structuralist and methodological realist accounts may appeal to posits about entities as well, which is not necessary for some versions of structural realism. For instance, Ladyman et al. 2007 defends a version of ontic structural realism without (necessary) commitment to scientific entities.

only in a limited domain surely are unsuccessful theories. Thus, there is no challenge to the usual arguments, and they may go on as before.

There is bad news for this response. General relativity is one of the most successful theories on the contemporary scene. But the field equations of general relativity are non-linear, and there is no general formula for their solution in a vast range of cases. Einstein's field equations, which are PDEs, determine how the spacetime metric tensor changes with respect to changes in massenergy.[15] Even when the equations are linearized, they cannot be solved directly in some of the most significant scientific contexts, famously including the inspiral and merger stages of a binary black hole ringdown.[16]

The Navier-Stokes equations[17] are partial differential equations describing the behavior of a fluid.[18] Contemporary applications describe the behavior of a fluid by solving the equations to find a velocity vector field. This results, usually, in a set of nonlinear PDEs, which are notoriously difficult to solve. John Wheeler has remarked on a kinship in this sense between the field equations of GR and the Navier-Stokes equations:

> An objection one hears raised against the general theory of relativity is that the equations are non-linear, and hence too difficult to correspond to reality. I would like to apply that argument to hydrodynamics-rivers cannot possibly flow in North America because the hydrodynamical equations are non-linear and hence much too difficult to correspond to nature! (Wheeler 2011, n.p.)

Both the field equations and the Navier-Stokes equations are sets of differential equations with deep grounding in physical laws.[19] Both include non-linear terms, and thus do not have straightforward solutions in the physical domains of interest: the strong field regime for GR, and most physical fluids for the Navier-Stokes equations.

Like the field equations, the original equations of fluid dynamics are not directly solvable in most physical contexts of interest. Wheeler's point was that, nonetheless, the Navier-Stokes equations are used daily in practical applications.[20] The field equations are used in practice, just as the Navier-Stokes equations are. But both

---

[15] Differential equations relate functions to their rates of change. Partial differential equations (PDEs) do so for multiple variables. The field equations of GR and the Navier-Stokes equations are both sets of partial differential equations.

[16] See Jamee Elder's paper for this volume (Elder 2023)

[17] Some refer to the Navier-Stokes "equation", and some to plural equations. It depends on how they are formulated.

[18] For detailed presentation of the equations, see McLarty 2023 and Darrigol 2002. For the "dispute over the viability of various theories of relativistic, dissipative fluids" see Curiel 2010.

[19] The Navier-Stokes equations are effectively formulations of conservation of mass and Newton's second law in the fluid domain. McLarty's paper for this volume (McLarty 2023) makes this point lucid.

[20] See Darrigol (Darrigol 2002, 95): "The Navier-Stokes equation is now regarded as the universal basis of fluid mechanics, no matter how complex and unpredictable the behavior of its solutions may be. It is also known to be the only hydrodynamic equation that is compatible with the isotropy and linearity of the stress-strain relation. Yet the early life of this equation was as fleeting as the foam on a wave crest. Navier's original proof of 1822 was not influential, and the equation was rediscovered or re-derived at least four times, by Cauchy in 1823, by Poisson in 1829, by Saint-Venant in 1837,

sets of equations lack direct solutions in many applied contexts of interest. Creative techniques including linearization, partial, constrained, or idealized solutions, and simulation are used to mediate between the equations and the natural phenomena being studied.

Summing up: Two widely accepted, empirically well founded theories, fluid dynamics and general relativity, rest on systems of partial differential equations that are not directly solvable in most domains of interest. These theories enjoy considerable experimental support and are able to generate effective explanations and predictions in their target regimes. We cannot account for their effectiveness by appealing to the ability to generate predictions and explanations directly from solutions to the equations in the primary domains of interest: the theories are not 'successful' in this sense. Thus, we need to explain how scientists and mathematicians can make progress with physical explanations in cases where reasoning based on the relevant equations alone is neither successful[21] nor complete.

We might marshal the no miracles argument to argue that the empirical success of these theories is best explained by the fact that they 'latch on' to real or objective relations in nature. Or we could formulate an empiricist account, explaining the theories' success by the contact they make with observable phenomena. Both of these accounts sound reasonable, but they float above the rough practical terrain they're supposed to explain. To really navigate this rocky terrain, we'd need to explain: How do theories of this kind make contact with the target phenomena?

One philosophical explanation of how theories reach out to objects emphasizes the role of models, conceiving them as neither the target phenomenon, nor part of the theory proper.[22] Models are constructed deliberately to build connections between theories and phenomena according to this account: the "models as mediators" view (Morrison 1999, Knuuttila 2005).[23] For models to work as mediators does not require that they be taken to be true or objective representations. Niels Bohr worked from the lines of the emission spectra of substances to construct a model involving discrete orbits of electrons around the nucleus. The development of the Bohr model is closely related to his formulation of an equation that correctly describes the relation between energy levels and spectral lines for the hydrogen atom (Bokulich 2011, Sect. 4). As Bokulich notes, the Bohr model is not an accurate depiction of the atom itself, but it allowed for the derivation of correct predictions via the formulation of equations. Bohr was able to relate observational data (emission lines and atomic

---

and by Stokes in 1845. ... All of these investigators wished to fill the gap they perceived between the rational fluid mechanics inherited from d'Alembert, Euler, and Lagrange, and the actual behavior of fluids in hydraulic or aerodynamic processes."

[21] In a very specific sense of 'successful': independently generating explanations and predictions.

[22] Some accounts allow for models to be objects, but I am focusing on those that see models as distinct: "what makes something a model explanation is that the explanans in question makes essential reference to a scientific model, and that scientific model (as I believe is the case with all models) involves a certain degree of idealization and/or fictionalization" (Bokulich 2011, 38).

[23] This account tacitly accepts a distinction between theory and observation, since otherwise there is nothing to 'mediate' (no gap to close). This distinction does not need to be as strong as that defended by some members of the Vienna Circle. It need only be made in practice.

spectral analysis) with theory (equations and predictions) by means of his atomic model.

Bokulich (2011) has proposed an account of 'structural model explanation', building on work by Morrison (1999) and Woodward (2003). Structural model explanations capture causal relations within target systems: a model "explains the explanandum by showing how the elements of the model correctly capture the pattern of counterfactual dependence of the target system. More precisely, in order for a model M to explain a given phenomenon P, we require that the counterfactual structure of M be isomorphic in the relevant respects to the counterfactual structure of P".[24] Once such relations of structural isomorphism are demonstrated to be valid, there is a further 'justificatory step', "specifying what the domain of applicability of the model is, and showing that the phenomenon in the real world to be explained falls within that domain" (39).

An idealized or even counterfactual model can be used to mediate between theory and target system, as long as (1) an isomorphism can be proven to exist between the relations of the model and the relations of the target system, and (2) it can be shown that the target system falls within the valid domain of application of the model. Leng (2021) observes that there is no reason structural model explanations of this kind could not apply to mathematical explanations within physical theories. Leng presents an account of "mathematical explanations as structural explanations", in which mathematical explanations "can be presented as deductively valid arguments whose premises include a mathematical theorem expressed modal structurally, together with empirical claims establishing that the conditions for the mathematical theorem are instantiated in the physical system under consideration" (Leng 2021, 10417).[25]

Leng's extension of the structural model explanation account to mathematical explanations is plausible. After all, Bokulich's account deliberately allows for highly idealized and even counterfactual explanations, as long as the required isomorphism between model and target system is achieved. But there are a few puzzling aspects of this extension prima facie. First, if mathematical explanations can be structural model explanations, then this pushes against the view that models are not part of a background theory. Many structural mathematical explanations are based on equations that are counted as part of a theory.

The 'models as mediators' view involve at least prima facie commitment to a gap between theories (which are idealized and abstract) and target systems (observable material things and their relations). If there is no gap, at least in practice, then there is nothing between which to mediate. Equations can, themselves, be regarded as models.[26] It may seem, then, as if we don't need anything beyond equations to

---

[24] Bokulich 2011, 39. A precedent to this account can be found in Heinrich Hertz's Bild theory; see (Eisenthal 2018).

[25] Importantly, Leng adds, "these explanations, which can be understood in modal structural terms, involve no commitment to mathematical objects platonistically construed" (Leng 2021, 10417).

[26] For discussion see (Sterrett 2023).

generate mathematical structural explanations in key cases, like the basic equations of many physical theories.

Another prominent account of models thus has it that they simply are equations, or, better, that some models are abstract systems that can ultimately be reduced to equations. For instance, according to this view the ideal gas law is not just an equation, but also a model from which can be derived predictions concerning the behavior of possible physical systems. More broadly, as Morrison notes, "We frequently refer to abstract formal systems as models that provide a way of setting a theory in a mathematical framework; gauge theory, for example, is a model for elementary particle physics insofar as it provides a mathematical structure for describing the interactions specified by the electroweak theory" (Morrison 2005, 145).

A system of partial differential equations could serve as a model in this sense. However, that would require being able to formulate relevant problems about the behavior of a system, and find solutions to those problems, in the domain of interest. Those solutions should be the source of structural descriptions that allow us to learn about the target systems.

4.3 Working to Formulate and Solve an Equation

An equation can provide "a mathematical structure for describing" the physical interactions in question, and thus can generate structural descriptions and explanations (Morrison 2005, 145). A structural mathematical explanation does not necessarily have to 'mediate' between theory and observation, then. A mathematical structure itself can be shown to be isomorphic to the relations of interest in the target system, and can be shown to be valid when applied to that system, thus satisfying Bokulich's criteria for structural explanation.[27]

It is particularly tempting to apply such an analysis to systems of partial differential equations. PDEs provide exactly the kind of causal, counterfactual explanation of the behavior of a system that grounds structural explanations. Systems of PDEs describe how one variable increases while another decreases, under precisely which conditions a system will remain at equilibrium, how a target system evolves over time, and so on.

Should we then conclude that the equations are the source of the relevant counterfactual, causal information, and of the resulting structural explanations? In that case, we wouldn't need a model separate from the equations. It is quite possible. But we need to think about why and how that might be. And, in particular, we need to consider what it takes to formulate, and then to solve, an equation or system of equations.[28]

---

[27] If the structural explanation is not in terms of a model distinct from theory, it does not meet Bokulich's criteria for a structural model explanation.

[28] This is a central theme of Susan Sterrett's contribution to this volume (Sterrett 2023).

In the case of the Navier-Stokes equations (and the field equations of GR), to reiterate: direct solutions are not available in many contexts of interest, because the original form of the equations is nonlinear. Nonlinear terms complicate the basic method for the solution of differential equations: finding the existence and uniqueness of solutions in a domain. Under certain conditions, given background constraints, the fluid or field equations can be reduced to linear PDEs. Many researchers use linearized forms of the equations to find solutions in a restricted domain.

Working with weak, regularized, or linearized forms of the equations can be a tool for learning about the potential for solutions in the strong or nonlinear cases. Jean Leray "was the first to study the Navier-Stokes equations in the context of weak solutions" (Ozanski and Pooley 2018, 114). Leray studied the equations on the "whole space R^3".[29] Leray's work allowed for the study of the equations from the perspective of weak solutions. By working with these solutions, he was able to derive "lower bounds on various norms of the strong solutions", and to "indicate the rate of blow-up" of strong solutions, among other results (Ozanski and Pooley 2018, 115). Leray's strategy was to work with a regularized form of the equations to derive weak solutions. Weak solutions become instruments for investigating the evolution of the equations in time and the possibility of achieving strong solutions in specific cases.[30]

Another strategy for learning about possible solutions to the Navier-Stokes equations is to generate solutions using simulations. Linear PDEs with boundary conditions, however complex, can be solved with a computer to yield physical solutions that predict the behavior of fluids (or of the metric, in GR). These solutions often require stipulating hypothetical values for key parameters, or generally making assumptions that allow for simulation of the target systems.

Using creative engineering, a simulation of a relevant class of phenomena might be provided.[31] The equations might then be shown to apply to the simulated phenomena.[32] If the right kinds of mediated connections between theory, simulation, and physical context could be found, then we can still argue that the equations are driving the structural explanations provided. One way to establish such a mediated connection would be to provide exact solutions to the equations for simulated situations.

---

[29] He studied the "regularized form" of the equations, "which are obtained by replacing the nonlinear term $(u \cdot \nabla)u$ by $J_\epsilon(u \cdot \nabla)u$ (where J_ϵ is the standard mollification operator)" (Ozanski and Pooley 2018, 115). A mollification operator is a smooth function applied to a nonsmooth one, which has the result of 'regularizing' or smoothing the function.

[30] McLarty 2023, Sects. 2.3-2.5 discusses weak solutions to Navier-Stokes.

[31] It is important to consider how engineering factors into heuristic reasoning. Sterrett has done significant work on reasoning about physically similar systems (Sterrett 2017a), analogue models (Sterrett 2017b), and dimensional analysis (Sterrett 2009).

[32] The LIGO detection of gravitational waves used simulations to excellent effect. See (Elder 2023).

4.3.1 Finding Exact Solutions to the Navier-Stokes Equations

This section examines a set of exact solutions to the Navier-Stokes equations that generate broader structural descriptions. Martin Bazant and Keith Moffatt have provided solutions that "describe steady vortex structures with two-dimensional symmetry in an infinite fluid" (Bazant and Moffatt 2005, 55). They describe "two classes of exact solutions of the Navier-Stokes equations":

1. The first is a class of similarity solutions obtained by conformal mapping of the Burgers vortex sheet to produce wavy sheets, stars, flowers and other vorticity patterns. (Bazant and Moffatt 2005, 55)

2. The second is a class of non-similarity solutions obtained by continuation and mapping of the classical solution to steady advection-diffusion around a finite circular absorber in a two-dimensional potential flow, resulting in more complicated vortex structures that we describe as avenues, fishbones, wheels, eyes and butterflies. These solutions exhibit a transition from 'clouds' to 'wakes' of vorticity in the transverse flow with increasing Reynolds number. (Bazant and Moffatt 2005, 55)

The similarity solutions are stable solutions, while the non-similarity solutions model systems that evolve beyond equilibrium with an increasing Reynolds number. The second class of solutions is particularly interesting for the insights it promises into vorticity.[33] Turbulence is characterized by increasing vorticity, and increasing turbulence results in the evolution of systems into nonlinear fluid systems. Thus, studying turbulence is central to understanding nonlinear fluid motion.

Bazant and Moffatt analyze vorticity beginning with a Burgers vortex. A steady "Burgers vortex sheet" is a vortex in which the diffusion of the vortex, its spreading outward and thus lessening, is counterbalanced exactly by convection (inward rotation) and stretching of the vortex.[34] In a Burgers vortex, diffusion and convection are balanced so that the vorticity of the fluid remains stable.[35] The formal relationship Burgers identified, emphasized here, turned out to be the basis of a number of exact solutions to the Navier-Stokes equations. These solutions in turn allowed for analysis of certain kinds of turbulence in terms of vortex shapes.[36]

---

[33] 'Vorticity' is informally described as the rotation of the fluid, and is often evaluated formally as the curl of the vector velocity field of the fluid.

[34] In a car wash, water is thrown outward so that it diffuses entirely. But when you pull the plug in a sink, the water is constrained by the shape of the sink and by the force that causes it to move inward. This is referred to as convection or advection (the latter if a material is transported by the fluid). A diagram is provided in (Jumars et al. 2009, 1).

[35] "In 1938 Taylor also recognised the fact that the competition between stretching and viscous diffusion of vorticity must be the mechanism controlling the dissipation of energy in turbulence. A decade later Burgers obtained exact solutions describing steady vortex tubes and layers in locally uniform straining flow where the two effects are in balance. Burgers introduced this vortex as 'a mathematical model illustrating the theory of turbulence', and he noted particularly that the vortex had the property that the rate of viscous dissipation per unit length of vortex was independent of viscosity in the limit of vanishing viscosity (i.e. high Reynolds number). The discovery of the exact solutions stimulated the development of the models of the dissipative scales of turbulence as random collections of vortex tubes and/or sheets" (Tryggeson 2007, 14).

[36] Usually cylinders, because Burgers gave his solution in cylindrical coordinates.

Bazant and Moffatt first obtain a class of similarity solutions via conformal mapping (Fig. 4.1). This mapping technique involves preserving local angles, but generating distinct solutions or maps.[37] They begin with a contour map of the steady Burgers vortex sheet (Bazant and Moffatt 2005, Fig. 4.1, 56). The initial class of solutions can be transformed back to that contour map. This is a class of distinct physical solutions, "self-similar" solutions generated by Möbius transformations. The solutions do not depend in any complex way on the Reynolds number, which is the ratio of inertial to viscous forces in the fluid. In fact, that dependence can be removed entirely for these solutions (Bazant and Moffatt 2005, 59).

Using these techniques, it is also possible to obtain a set of solutions based on a dynamic process. These solutions involve generating "avenues" through a fluid by solving the "canonical problem of steady advection-diffusion around an absorbing circular cylinder in a uniform background flow of constant velocity and constant concentration" (Bazant and Moffatt 2005, 59). In other words, a cylinder that can take on water is dragged through a uniformly flowing and uniformly dense background fluid. The cross-section of the results is displayed in Fig. 4.2. [38]

Once these avenues have been generated, conformal mapping allows for the generation of non-self-similar but exact solutions to the Navier-Stokes equations (Fig. 4.3).

These are a set of vortex structures or 'avenues', described as 'fishbones, wheels, eyes, and butterflies', that constitute dynamical solutions to the equations in simulated situations. Bazant and Moffatt thus obtain a different set of solutions, generated by applying conformal mapping to vortex avenues. Vortex avenues have a "nontrivial dependence" on the Reynolds number, which is a dimensionless parameter that depends on the viscosity of the fluid. The second set of solutions cannot be mapped directly back on to the Burgers vortex sheet. They are built from simulations of vortex avenues that are complicated by increasing vorticity (possibly, turbulence), which is not a symmetrical evolution of the fluid system.

The authors conclude,

> We have presented two classes of exact solutions of the Navier-Stokes equations representing steady vortex structures with two-dimensional symmetry, confined by transverse potential flows. These solutions provide mathematical insights into the Navier-Stokes equations and physical insights into ways that vorticity may be confined. They also provide stringent tests for the accuracy of numerical simulations. (Bazant and Moffatt 2005, 63)

From the perspective of the account of mathematical structural explanation discussed above, these results are intriguing. The solutions derived by Bazant and Moffatt construct "steady vortex structures" that allow for descriptions of the strucural properties of fluid systems and their evolution over time, and insights into the mathematics involved. Moreover, the vortex avenues generated by Bazant's and Moffatt's methods are based on simulated systems. The solutions they provide are thus valid for those systems, which is how it can be proven that the solutions apply to physical phenomena.

---

[37] See Sterrett 2023 for detailed analysis.

[38] For all figures, the captions are reproduced from the original.

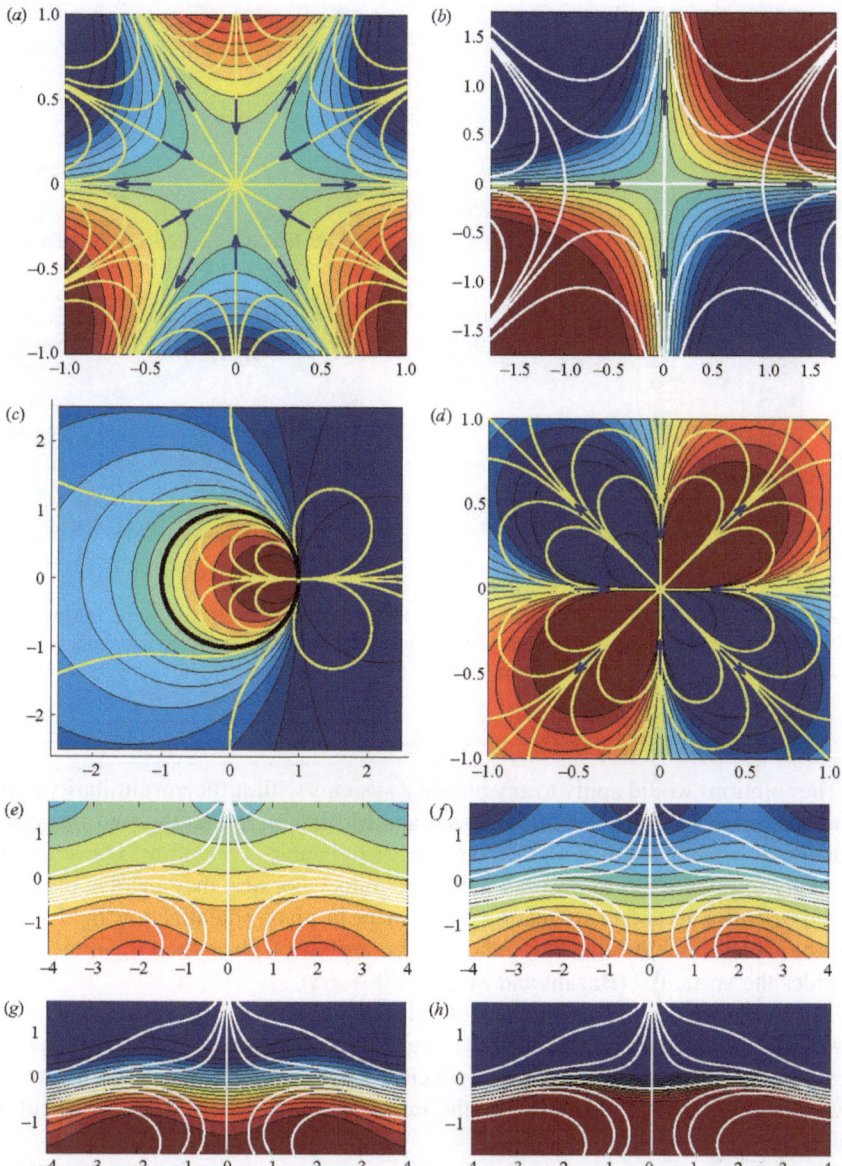

Fig. 4.1 Solutions obtained by conformal mapping, $\omega = f(z) = z^3$; (b) a vortex cross with stagnation points on its arms, $f(z) = 1-z^2$;(c) a circular vortex sheet $f(z) = (1 + z)/i(1-z)$ (where the black circle indicates the separatrix of the transverse flow); (d) a vortex flower, $f(z) = z^{-2}$; and(e-h) wavy vortex sheets, non-uniformaly starined by dipole singularities $f(z) = z + \sum_{j=1}^{3} (z - z_j)^{-1}, z_1 = 2.5i, z_2 = 2.5i, z_3 = 2-2.5i,$for(e) $Re = 0.01,(f)$ $Re = 0.1,(g)Re = 1,$and(h)$Re = 10$

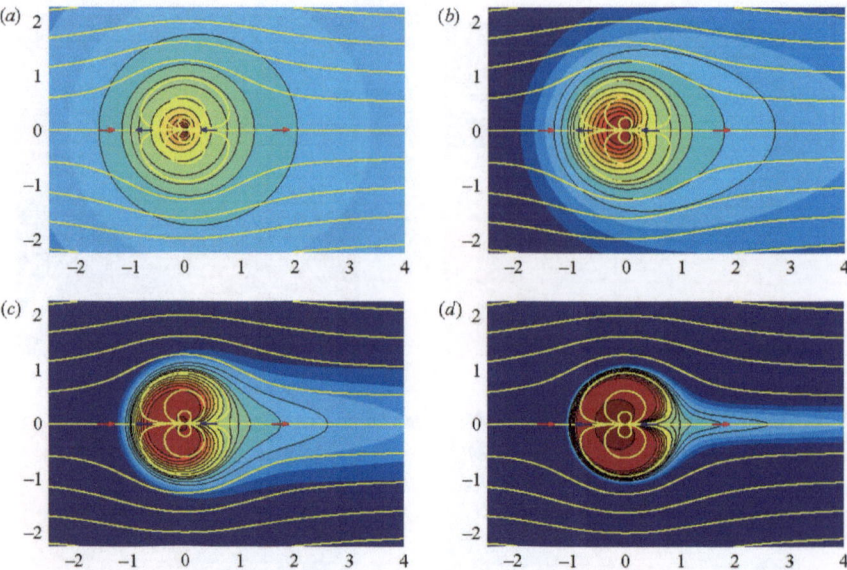

Fig. 4.2 Steady vortex avenues confined by transverse flow with a dipole source imside and a unifrom background (a dipole at ∞). These transverse jrts exhibit a non-trivial dependence on the Revnolds number with a transition from 'clouds' to 'wakes' of vorticity $(a) Re = 0.1$, $(b) Re = 1$, $(c) Re = 10$, $(d) Re = 100$

The solutions would apply to any physical systems with sufficient similarity to the simulated ones they employ. Still, Bazant and Moffatt warn that the set of physical situations to which these exact solutions will be applicable is very thin: "Our solutions are likely to be difficult to observe in the laboratory because they involve carefully placed singularities and/or stagnation points in the in-plane potential flow field, which must be achieved while also setting the appropriate out-of-plane shearing flow, which provides the vorticity" (Bazant and Moffatt 2005, 63).

The fluid flows in the second set of solutions depend on jets of water that are flowing in different directions (arrows mark this). The avenues generated involve carefully placing hypothetical jets of water in ways that stabilize turbulence of the flow. Placing the jets and stabilizing the turbulence will be much more difficult in reality than in a simulated situation.

The generation of exact solutions to the equations in these simulated situations validates the solutions for physical fluids, however hypothetical. This accomplishes one of Bokulich's criteria for structural explanation, that the structures generated must be shown to apply to the physical domain of interest. In fact, Bazant's and Moffatt's solutions meet all of Bokulich's requirements for structural explanations. They allow for the articulation of fluid systems as structures (vortex avenues) that can be shown to be isomorphic to physical target systems, and they can be shown to validly apply to such systems. While the solutions are illustrated with diagrams

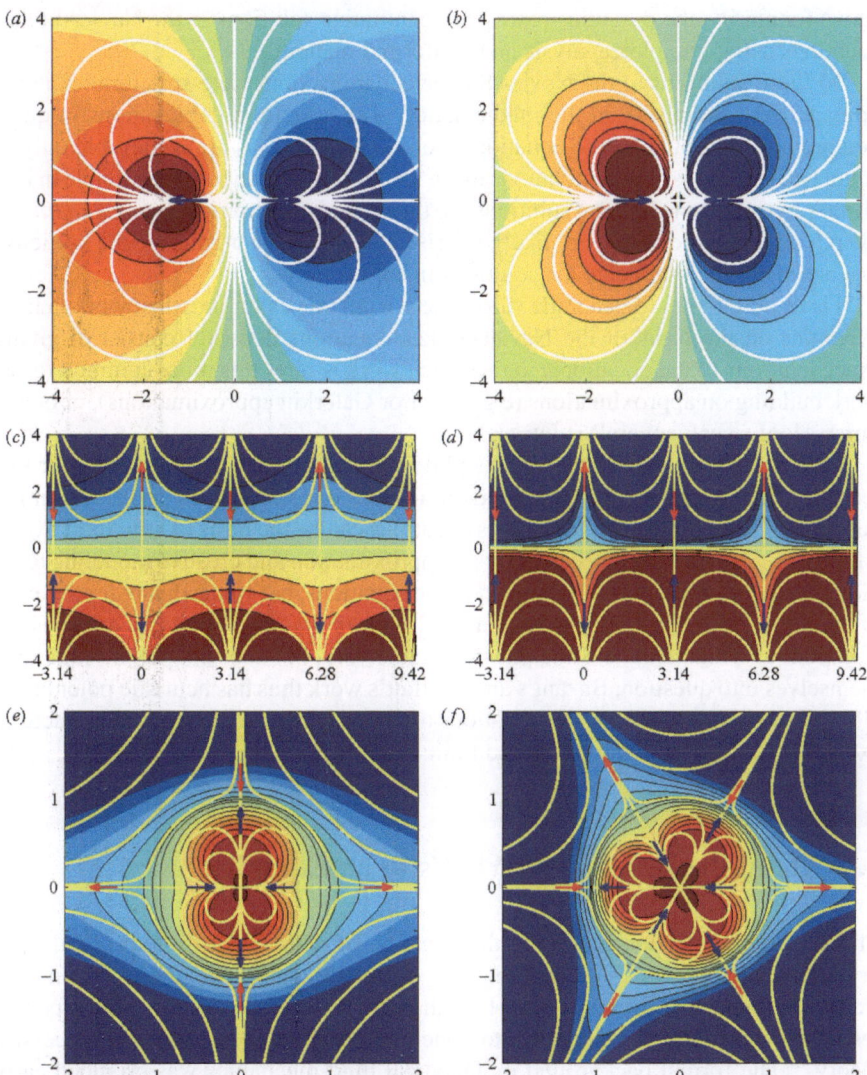

Fig. 4.3 Solutions obtained by conformal mapping, $\omega = f(z)$, to the vortex avenues in Fig. 4.3: voretex eyes, $f(z) = (1+z)/(1-z)$ a pair of avenues stabilized by transverse diverting dipoles at $Re = 10$; (b) a vortex butterfly, $(f) = i\log_z$, at (c) $Re = 0.1$ and (d) $Re = 10$; and vortex wheels, (e) $f(z) = z^2$ and (f) $f(z) = z^3$, at $Re = 1$

above for clarity, the diagrams represent quantitative solutions to the Navier-Stokes equations using complex analysis and conformal mapping.

Do these solutions to the Navier-Stokes equations constitute structural explanations derived directly from the mathematics? The structural explanations in question are not generated by the equations alone. The solutions are achievable only in simulated situations. The simulations make solutions to the equations possible in idealized physical contexts. The solutions are thus based on empirical *understanding* of the physical situations that are likely to arise, and of the phenomena (like vorticity) that are likely to complicate the solutions in applications.

It is important not to generalize too widely from this example. Most work in fluid dynamics on the basis of the Navier-Stokes equations does not consist of giving exact solutions to the equations: in fact, such work is rare. It is more common to see work building on approximations (e.g. linear or Galerkin approximations), or on the Leray-Hopf (weak general) solutions.

But Bazant's and Moffatt's work is philosophically interesting precisely for this reason. Much of the technical work done in fluid dynamics does not touch on the question of whether the Navier-Stokes equations can be the basis of structural explanations. Instead, much current work uses approximation and creative modeling techniques to solve problems in the context of the equations, but somewhat independently of them. By "independent of the equations", I mean that the work done assumes the equations as a mathematical and physical framework, but does not put the equations themselves into question. Bazant's and Moffatt's work thus has heuristic potential: it allows us to investigate how the equations themselves can be the source of structural explanations, and what the scope and limits of those explanations can be.[39]

4.4   Heuristic Reasoning: What's In an Equation?

This concluding section returns to the question of what makes a 'successful' theory. The development of fluid dynamics, from the earlier work of D'Alembert, Euler, Lagrange, the Bernoullis, and Helmholtz to the contributions of contemporary researchers, provides every reason to conclude that fluid dynamics is a successful theory.[40] The formal background in analytical fluid mechanics was provided early on. Its extension to modeling was provided by the partial differential equation formu-

---

[39] More work in fluid dynamics could be analyzed using this framework, of course.

[40] As Olivier Darrigol notes, the origins of the Navier-Stokes equations were in the attempt to mediate between fluid mechanics and actual fluid behavior: "All of these investigators wished to fill the gap they perceived between the rational fluid mechanics inherited from d'Alembert, Euler, and Lagrange, and the actual behavior of fluids in hydraulic or aerodynamic processes" (Darrigol 2002, 95).

lation by Navier and Stokes.[41] The contemporary theory of fluids is used in multiple contexts to achieve practical and theoretical results.

The theory of fluid dynamics is not 'successful' in one traditional sense: it does not support the derivation of predictions and explanations in some relevant domains directly from the equations of the theory.[42] But that traditional sense of 'success' is much too limited. It focuses on one method: deriving predictions, representations, and results from the equations in isolation. A more accurate picture of the use of equations would also take into account another process: how formulations of and solutions to equations build on scientists' understanding of empirical and simulated situations to generate structural explanations.

I would call this a 'heuristic' approach to scientific explanation.[43] The term 'heuristic' is sometimes used in a narrow sense, e.g., in decision theory, to describe mental shortcuts for solving problems. That is very far from the sense I want to identify. Instead, I want to use the sense of 'heuristic' that Einstein used when he called the principle of covariance a 'heuristic aid': structural reasoning that informs the search for scientific explanations and the assessment of whether the phenomena are in agreement with a theory, especially in novel experimental contexts.[44]

Elaborating this stronger sense of heuristic reasoning allows for a better understanding of how equations function in the development of scientific explanations over time. Given the account of structural explanation elaborated above, one might conclude that there are two ways to account for the role of equations. We might argue that equations cannot generate structural explanations on their own: that independent models are needed to mediate between abstract 'background' theories and physical observable phenomena. On the other hand, we might argue that abstract equations

---

[41] The particular techniques used here owe a debt to Burgers, who identified a number of formal relationships that do not depend on the Reynolds number (indexed to the viscosity of the fluid), which allowed for leaps forward in the modeling of fluids.

[42] This is the sense of scientific explanation corresponding to the Deductive-Nomological Model.

[43] There are heuristic accounts of scientific models which are clearly related, but modeling is not the focus here.

[44] "The law of transmission of light, the acceptance of which is justified by our actual knowledge, played an important part in this process of thought. Once in possession of the Lorentz transformation, however, we can combine this with the Principle of Relativity, and sum up the theory [...] in brief: General laws of nature are co-variant with respect to Lorentz transformations. This is a definite mathematical condition that the theory of relativity demands of a natural law, and in virtue of this, the theory becomes a valuable heuristic aid in the search for general laws of nature. If a general law of nature were to be found which did not satisfy this condition, then at least one of the fundamental assumptions of the theory would have been disproved" (Einstein 1921, 50–51). A treatment of the heuristic role of mathematics in relativity, as well as how Einstein's heuristic reasoning developed over time, is found in section 3 of Zahar 1980. Recently, Christopher Pincock has traced the derivation of Poiseuille's law and explicitly analyzes "how an experimental discovery can prompt the search for a theoretical explanation and also how obtaining such an explanation can provide heuristic benefits for further experimental discoveries" (Pincock 2021, 11667). Pincock's paper appears in a topical collection edited by Sorin Bangu, Emiliano Ippoliti, and Marianna Antonutti, "Explanatory and Heuristic Power of Mathematics" (Bangu et al. 2021). An earlier collection focuses on heuristic reasoning in particular (Ippoliti 2015).

on their own yield explanations in physical domains of interest directly, without any mediation needed.

The third account that I want to articulate is a heuristic account of structural explanation.[45] Solutions to equations can yield structural explanations that have epistemic, heuristic value even before they have been experimentally tested. These solutions need not be derived from the equations in isolation, however. Rather, they involve (and support) significant understanding of the empirical, experimental, and mathematical situation in which the equations were formulated. That understanding may be mobilized through simulation of the relevant phenomena-for instance, by the use of hypothetical general cases. In some situations, the necessary framework for solutions is achieved by models used as mediators. But in other situations, the techniques may not involve models, and here we need a broader account.

First, the movement between data (e.g., measurements) and the formulation of the relevant problem is far from trivial. In the context of differential equations, the problem, as Curiel has noted, is how to characterize "the transitions to and fro between, on the one hand, inaccurate and finitely determined measurements, and, on the other, the mathematically rigorous initial-value formulation of a system of partial-differential equations" (Curiel 2010, Sect. 1). Once an initial value problem has been formulated and initial and boundary conditions found, it is more straightforward to find a solution.[46]

Heuristic solutions like Bazant and Moffatt's can play the role of epistemic light-houses. They define, not just classes of models associated with a theory, but broader depictions of how the theory could be instantiated: how the equations of the theory could be solved in well classified physical situations. Even if the physical solutions thus achieved are unlikely - e.g., scientists do not expect to encounter the exact physical situations in nature-they provide standards or benchmarks that give working scientists pointers about how to proceed and how to evaluate the results they do achieve.[47] A key heuristic value of simulated solutions is to set standards for actual physical solutions. Bazant's and Moffatt's results provide exactly such a check, for a broad domain of physically meaningful situations.[48]

---

[45] Existing accounts of heuristic explanation include Einstein 1921, Eisenthal 2018, Ippoliti 2015, and Obradović et al. 2009. Pincock 2021 deserves particular mention, since Pincock here develops a heuristic interpretation of the derivation of Poiseuille's law in fluid mechanics.

[46] Solving the initial value problem alone, however, does not prove a solution applies everywhere in a domain. In fact, that solution process does the opposite: it involves precisely determining how to find a solution to the equation in a specific physical situation. For instance, a solution or set of solutions may split up the domain into simulated cases, as Bazant and Moffatt do in the paper examined above. Simulations have multiple roles, including providing methods for data assimilation and ways to relate measurements to equations (Parker 2017).

[47] As Sterrett observes, simulations can extend reasoning to physically similar or analogue systems (Sterrett 2017a, Sterrett 2017b, Sterrett 2023).

[48] Bazant and Moffatt's solutions can even be regarded as epistemic 'artifacts', and thus are potentially consistent with the artifactual account of scientific modeling articulated in Knuuttila 2021. However, Knuuttila emphasizes the independence of artifactual models from 'real-world systems' (Knuuttila 2021, S5079). The heuristic account emphasizes isomorphism between structural explanations and real-world systems.

The heuristic account of structural explanation provides an explanation of how scientific theories can be successful, and yet the equations of those theories not be solvable in target domains of the theory. The equations can be given simulated, idealized, or limited solutions that are the basis of structural explanations in those domains. Structural isomorphism between the simulated or idealized solutions to the equations and the real-world systems in question yields structural explanations that work well in practice.

It is key, though, that the process of finding solutions to equations in specific physical contexts - even if those contexts are simulated - is not reducible to deriving predictions based on the abstract equations alone. Finding a solution to an equation in a particular context may involve substantial empirical and experimental reasoning, as well as creative mathematical reasoning, as Sterrett (2023) argues convincingly. The point is made well by Erik Curiel.

> one also needs all the collateral knowledge, both theoretical and practical, not contained in the equation, in order to apply the equation to the modeling and comprehension of all the phenomena putatively treated by the theory. To put the matter more vividly: the equation as a result of a (profound) investigation of the physical phenomena at issue, of all the empirical data and attempts to model that data heretofore, a teasing apart and characterization of the maximally common structure underlying the system of relations that obtains among them (supposing there is such a thing)-the sort of investigation that Newton and Maxwell and Einstein did and Hilbert did not accomplish-the equation as a result of that is the theory. (Curiel 2010 (preprint), Sect. 7)

Equations are the result of substantial structural reasoning, built up over time and involving empirical and experimental investigation. Abstract equations can yield predictions and explanations, even when they can't be directly solved in a specific domain. But it is crucial, in such cases, to scrutinize what it takes to find simulated, limited, or idealized solutions.[49] That scrutiny may reveal limitations to the scope of the equations. The creative process of finding solutions in such cases can provide substantial heuristic information about the value of the equations and the strength of the theory.

References

Albritton D, Bruè E, Colombo M (2022) Non-uniqueness of Leray solutions of the forced Navier-Stokes equations. Ann Math 196(1):20220701

Bangu S, Ippoliti E, Antonutti M (2021) Explanatory and heuristic power of mathematics. Top Collect Synth

Bazant M, Moffatt K (2005) Exact solutions of the Navier-Stokes equations having steady vortex structures. J Fluid Mech 541:55–64

Bokulich A (2017) Models and explanation. In: Magnani L, Bertolotti T (eds) Handbook of model-based science. Springer, Dordrecht, pp 103–118

Bokulich A (2011) How scientific models can explain. Synth 180:33–45

---

[49] The recent nonuniqueness results by Albritton, Bruè, and Colombo (Albritton et al. 2022) are an excellent case of close scrutiny of known solutions.

Boyd N (2018) Evidence enriched. Philos Sci 85(3):403–421

Chang H (2009) We have never been whiggish (about phlogiston). Centaur 51:239–64

Colyvan M (2015) Indispensability arguments in the philosophy of mathematics. In: Zalta EN (ed.) The stanford encyclopedia of philosophy. https://plato.stanford.edu/archives/spr2015/entries/mathphil-indis/

Cordero A (2011) Scientific realism and the divide et impera strategy. Philos Sci 78(5):1120–1130

Curiel E (2016) Kinematics, dynamics, and the structure of physical theory. Preprint at https://doi.org/10.48550/arXiv.1603.02999

Curiel E (2010) On the formal consistency of theory and experiment, with applications to problems in the initial-value formulation of the partial-differential equations of mathematical physics. Preprint at http://philsci-archive.pitt.edu/8660/

Darrigol O (2002) Between hydrodynamics and elasticity theory: the first five births of the Navier-Stokes equation. Arch Hist Exact Sci 56(2):95–150

Einstein A (1921) Relativity: the special and the general theory. Trans. and ed. Lawson R. Henry Holt, New York

Eisenthal J (2018) Mechanics without mechanisms. Stud Hist Philos Sci Part B 62:45–55

Elder J (2023) Black hole coalescence: observation and model validation. In: Working toward solutions in fluid dynamics and astronomy: what the equations don't say. Springer Briefs

Ippoliti E (ed) (2015) Heuristic reasoning. Springer, Cham

Jumars P, Trowbridge J, Boss E, Karp-Boss L (2009) Turbulence-plankton interactions. Mar Ecol 30(2):133–50

Kitcher P (1993) The advancement of science. Oxford University Press, Oxford

Knuuttila T (2005) Models, representation, and mediation. Philos Sci 72(5):1260–1271

Knuuttila T (2021) Imagination extended and embedded: artifactual versus fictional accounts of models. Synth 198(Supplement 21):S5077–S5097

Ladyman J, Ross D, Spurrett D, John C (2007) Every thing must go. Oxford University Press, Oxford

Leng M (2021) Models, structures, and the explanatory role of mathematics in empirical science. Synth 199:10415–10440

Leplin J (1986) Methodological realism and scientific rationality. Philos Sci 53(1):31–51

Lyons T (2006) Scientific realism and the stratagema de divide et impera. Br J Philos Sci 57(3):537–560

McLarty C (2023) Fluid mechanics for philosophers, or which solutions do you want for Navier-Stokes?. In: working toward solutions in fluid dynamics and astronomy: what the equations don't say. Springer Briefs

Morrison M (1999) Models as autonomous agents. In: Morgan M, Morrison M (eds) Models as mediators 1999. Cambridge University Press, Cambridge, pp 38-65

Morrison M (2005) Approximating the real: the role of idealizations in physical theory. In: Idealization XII: Correcting the model. Brill, Leiden

Obradović S, Ninković S (2009) The heuristic function of mathematics in physics and astronomy. Found Sci 14(4):351–360

Ozanski W and Pooley B (2018) Leray's fundamental work on the Navier-Stokes equations. In: Fefferman, Robinson, and Rodrigo (eds) Partial differential equations in fluid mechanics. Cambridge University Press, Cambridge

Parker W (2017) Computer simulation, measurement, and data assimilation. Br J Philos Sci 68(1):273–304

Patton L (2015) Methodological realism and modal resourcefulness. Synth 192(11):3443–63

Pincock C (2021) The derivation of Poiseuille's law: heuristic and explanatory considerations. Synth 199:11667–11687

Psillos S (1999) Scientific realism: how science tracks truth. Routledge, London

Psillos S (2018) Realism and theory change in science. In: Zalta E (ed) The stanford encyclopedia of philosophy.

Quine WvO (1950) Methods of logic, 1st edn. Holt

Quine WvO, Ullian JS (1978) The web of belief, 2nd edn. McGraw-Hill, New York

Resnik M (1995) Scientific vs. mathematical realism. Philos Math 3(2):166–174

Sterrett S (2009) Similarity and dimensional analysis. In: Mejers A (ed) Philosophy of technology and engineering sciences, Elsevier, Amsterdam, pp 799–823

Sterrett S (2017a) Physically similar systems - a history of the concept. In: Magnani L, Bertolotti T (eds) Springer handbook of model-based science. Springer, Switzerland

Sterrett S (2017b) Experimentation on analogue models. In: Magnani and Bertolotti (eds)

Sterrett S (2023) How mathematics figures differently in exact solutions, simulations, and physical models. In: Working toward solutions in fluid dynamics and astronomy: what the equations don't say. Springer Briefs

Tryggeson H (2007) Analytical vortex solutions to the Navier-Stokes equation. Växjö University Press

Wheeler J (2011) The present position of classical relativity theory and some of its problems. The role of gravitation in physics. Max-Planck-Gesellschaft, Berlin

Worrall J (1989) Structural realism: the best of both worlds? Dialectica 43(1–2):99–124

Zahar E (1980) Einstein, Meyerson and the role of mathematics in physical discovery. Br J Philos Sci 31(1):1–43

Chapter 5
Black Hole Coalescence: Observation and Model Validation

Jamee Elder

Abstract This paper will discuss the recent LIGO-Virgo observations of gravitational waves and the binary black hole mergers that produce them. These observations rely on having prior knowledge of the dynamical behaviour of binary black hole systems, as governed by the Einstein Field Equations (EFEs). However, we currently lack any exact, analytic solutions to the EFEs describing such systems. In the absence of such solutions, a range of modelling approaches are used to mediate between the dynamical equations and the experimental data. Models based on post-Newtonian approximation, the effective one-body formalism, and numerical relativity simulations (and combinations of these) bridge the gap between theory and observations and make the LIGO-Virgo experiments possible. In particular, this paper will consider how such models are validated as accurate descriptions of real-world binary black hole mergers (and the resulting gravitational waves) in the face of an epistemic circularity problem: the validity of these models must be assumed to justify claims about gravitational wave sources, but this validity can only be established based on these same observations.

Keywords Black holes · Gravitational waves · General relativity · Models · Simulations · Astrophysics

5.1 Introduction

On September 14, 2015 the "Laser Interferometer Gravitational-wave Observatory", (hereafter "LIGO"), comprising interferometers in Hanford, WA, and Livingston, LA, detected gravitational waves for the first time. This event, dubbed GW150914 (based on the date of detection), marked the beginning of a new epoch for the field of gravitational-wave astrophysics. The two Advanced LIGO interferometers, together

J. Elder (✉)
Black Hole Initiative, Harvard University, 20 Garden Street, 2nd Floor,
Cambridge, MA 02138, UK
e-mail: jelder@fas.harvard.edu

© The Author(s), under exclusive license to Springer Nature Switzerland AG 2023
L. Patton and E. Curiel (eds.), *Working Toward Solutions in Fluid Dynamics and Astrophysics*, SpringerBriefs in History of Science and Technology,
https://doi.org/10.1007/978-3-031-25686-8_5

with the Advanced Virgo observatory in Italy, form a global network of gravitational-wave interferometers capable of observing gravitational waves and (through these) the astrophysical sources that generate them.

The detection of gravitational waves has been hailed as a revolution for astrophysics. This is because the recent advent of gravitational-wave astrophysics gives us a new "window" through which to observe the universe, enabling us to observe events and objects that were previously invisible to us, such as collisions between two black holes. Black holes and gravitational waves are gravitational phenomena predicted by Einstein's theory of general relativity. On the standard geometric interpretation of this theory, gravity is viewed as a manifestation of the curvature of spacetime. This curvature is encoded in the (Lorentzian) metric, $g_{\mu\nu}$, and the response of the metric to energy and momentum is governed by the Einstein field equations,

$$G_{\mu\nu} \equiv R_{\mu\nu} - \frac{1}{2}g_{\mu\nu}R = 8\pi T_{\mu\nu} \qquad (5.1)$$

The right hand side of this equation concerns the local energy and momentum within the spacetime, as expressed by the energy-momentum tensor (alternatively called the stress-energy tensor) $T_{\mu\nu}$. The left hand side concerns the curvature of spacetime. Here, $G_{\mu\nu}$ is the Einstein tensor, defined in terms of the Ricci tensor, $R_{\mu\nu}$, the metric, $g_{\mu\nu}$, and the Ricci scalar, R.

Equation (5.1) can be viewed as an equation relating 4×4 matrices. The components of these matrices yield a total of sixteen non-linear partial differential equations for the metric $g_{\mu\nu}$. However, the symmetry of the metric tensor means that this reduces to only 10 independent equations.[1]

Unfortunately, exact solutions to the Einstein field equations are difficult to come by. While many such solutions now exist, they often describe very simple, idealised physical scenarios. Much of the usefulness of such solutions is in providing a starting point for approximation schemes, used to model more physically realistic systems (Kennefick 2007, 41). Such approximation schemes can even be used even when there are no exact solutions to the Einstein equations to use as a foundation. Instead, empirically successful descriptions provided by previous theories are used. For example, Einstein's famous calculation of Mercury's perihelion advance started from the Newtonian solution then added relativistic corrections in powers of $(v/c)^2$ (Kennefick 2007, 42–43). This "post-Newtonian" approximation scheme remains in use today. Indeed, compact binaries are well-modelled by post-Newtonian approximation for the early inspiral, where the characteristic velocities and the gravitational field strength remain small (in relativistic terms). Beyond the early inspiral the compact binary merger must be described in the "dynamical strong field regime", where velocities are high and the gravitational field is strong. Here, the lack of full analytic solutions becomes more problematic, as I will discuss in the remainder of this paper.

---

[1] See Kennefick (2007, 46–47) for a very clear explanation of this point.

Despite the lack of exact analytic solutions, recent work has led to a range of approaches to modelling the late stages of the merger, notably including the effective one-body formalism and numerical relativity simulations. This paper concerns the validation of such models in the absence of exact analytic solutions.[2]

Interwoven with questions about the validity of these models are questions about the status of the LIGO-Virgo observations. These observations rely heavily on the use of such models for both gravitational wave detection (through matched filtering, see Sect. 5.2.1) and inferences about the compact binary systems that produced them (through parameter estimation, see Sect. 5.2.2). Indeed, advances in numerical relativity and other modelling approaches were crucial to the success of LIGO-Virgo experiments.

In this paper, I argue that modelling plays an essential role in connecting high-level theory, embodied in the Einstein field equations, with the LIGO-Virgo data. The models used in template-based searches for gravitational waves (Sect. 5.2.1) and in parameter estimation (Sect. 5.2.2) incorporate insights from a range of modelling approaches, allowing them to bridge the gap between theory and world. Thus, alongside technological advances, the modelling approaches discussed play a vital role in gaining empirical access to both gravitational waves and the binary black hole mergers that produce them.

However, validating these models—both as representative of the predictions of general relativity, and as accurate descriptions of the target systems—presents a challenge. This is because the model-dependence (or theory-ladenness) of the LIGO-Virgo observations creates a justificatory circularity: the accuracy of the models must be established using the LIGO-Virgo observations, but these observations *assume* the accuracy of the models. Additional features of the epistemic situation, such as the lack of independent access to these systems and the new physical regimes being probed, render this circularity difficult to break. However, the LIGO-Virgo Collaboration does perform "tests of general relativity" that test specific assumptions in the experimental methodology. While this doesn't exactly break the circularity, it goes some way towards rendering it benign, or even virtuous.

Overall, this paper shows how the methodology of the LIGO-Virgo experiments is intimately bound up with models of binary black hole mergers and the gravitational waves that they produce. The success of these experiments rests on confidence in these models, which bridge the gap between theory and phenomena. The flip side of this is that much of the interesting work in validating the LIGO-Virgo results lies in validating the models themselves with respect to both the equations of general relativity and the physical systems being observed.

---

[2] There are distinctions between solutions that are "exact", "elementary", "algebraic" etc., which I do not delve into here. For discussion of these issues, see Fillion and Bangu (2015).

5.2   Modelling and Observing Binary Black Hole Mergers

The LIGO and Virgo interferometers produce gravitational-wave strain data as a time series, sampled more than 16,000 times per second. This data is very noisy, with terrestrial noise sources disguising even strong signals like GW150914. This means that it is not possible to see a signal in the unprocessed data. Instead, significant data-processing is needed both to search for gravitational wave signals in the data and to infer the properties of the systems that produced them.

5.2.1   Template-Based Searches for Gravitational Waves

The LIGO-Virgo Collaboration has multiple independent search pipelines that are used to find gravitational waves in their data. This includes unmodelled "burst" searches using, for example, the coherent Waveburst ("cWB") algorithm. However, the most effective search methods are modelled searches, which rely on assumptions about the kinds of signals being sought. In this section, I'll mostly be concerned with reviewing the most important feature of these modelled searches: matched filtering. For more comprehensive descriptions of the data analysis techniques used, see, for example, Abbott et al. (2016a, 2020).

Matched filtering is a signal-processing technique that involves correlating a known signal, or "template", with an unknown signal, in order to detect the presence of the template within the unknown signal.[3] This technique allows for the extraction of a pre-determined signal from much larger noise. An optimal filtering maximises the "signal-to-noise ratio" (hereafter "SNR"), which (roughly) measures the ratio of the (amplitude of the) gravitational wave signal compared to the noise.

In order to extract a particular gravitational wave signal, the data must be searched using a closely matching template. Since it isn't known in advance which (if any) signal is present, the data must be searched for a range of signals, corresponding to the full range of gravitational waves that might be present. What is needed, then, is a library of template models corresponding to the range of possible signals. Templates are arrived at by considering the range of systems that we believe could produce measurable gravitational waves and determining the predictions of general relativity for the behaviour of such systems, including gravitational wave emission. Of course, this is no trivial task (to put it mildly).

Modelled searches using matched filtering appear to be a clear case of theory-laden observation. What can be "seen" in the data is determined by the models that are used to in the observation process. The theory- or model-ladenness of these observations naturally leads to concerns about whether the LIGO-Virgo Collaboration are observing all of and only the genuine gravitational wave signals in the data, and how accurate these observations are (i.e., how well the recovered signal reflects the grav-

---

[3] For a detailed introduction to matched-filtering in gravitational-wave astrophysics, see (Maggiore 2008, Sect. 7.3).

itational waves passing through the interferometer). I discuss the potential pitfalls of the model-based searches in Sect. 5.3.1.

5.2.2   Bayesian Parameter Estimation

Having detected a gravitational wave signal, it is then possible to make inferences about the properties of the source system that produced it (a compact binary merger, usually a binary black hole merger). This involves estimating the values of a range of parameters characterising the system. Through parameter estimation, the detection of gravitational waves doubles as an observation of a compact binary merger (at least on permissive uses of the term "observation", see e.g., Shapere (1982) and Elder (2020, Chap. 2). Indeed, GW150914 was called the *first* observation of a binary black hole merger.

There are 15 key parameters that determine the received signal.[4] These are:

- Luminosity distance to coalescence event (1 parameter)
- Angular location of event in sky (2 parameters)
- Orientation of the orbital plane relative to line of sight (2 parameters)
- Time of arrival (1 parameter)
- Orbital phase at time t (1 parameter)
- Masses of the component compact objects (2 parameters)
- Spin components of compact objects (6 parameters, 3 per object).

Of these, only the last two list items are "intrinsic parameters"; they concern the properties of the compact binary itself rather than its relationship to the interferometers used to observe it.

Parameter estimation is performed within a Bayesian framework. The basic idea is to calculate posterior probability distributions for the parameters describing the source system, using Bayes' theorem. To do this, we need the following: a model M that takes a set of system parameters and predicts the resulting signal; background or prior information I; and some data d. Bayes' theorem can then be written as:

$$p(\theta|d, M, I) = p(\theta|M, I)\frac{p(d|\theta, M, I)}{p(d|M, I)}, \qquad (5.2)$$

where $\theta = \{\theta_1, \ldots \theta_N\}$ is a collection of parameters. This equation tells us how to calculate the posterior distributions for θ, given some fixed modelling approach M and other analysis assumptions I.[5]

---

[4] This assumes that the eccentricity of the orbit can be neglected, since the emission of gravitational radiation is expected to circularise the orbit by the time the emitted gravitational waves enter the bandwidth of the detector (Peters 1964).

[5] For a details, see Abbott et al. (2020, 36).

On this approach, prior probability distributions must be specified for all 15 parameters. For some parameters this can be done based on symmetries of the parameter space. For example, for a redshift $z \ll 1$ in a Friedmann-Lemaître-Robertson-Walker cosmological model, equal numbers of coalescence events are expected to occur in equal co-moving volumes (Abbott et al. 2020, 37). For other parameters, the LIGO-Virgo approach is to choose simple priors such that the posteriors can be easily interpreted.

In addition to the parameters characterising properties of the binary, this analysis also includes $\mathcal{O}(10)$ extra parameters per detector to model calibration uncertainties. Thus for a three-detector analysis around 45 parameters are being sampled. In order to efficiently sample this high-dimensional parameter space, the LIGO-Virgo Collaboration developed LALInference, a stochastic sampling library that uses two different algorithms to perform parameter estimation. The first is a parallel tempering Markov chain Monte Carlo algorithm, and the second is a nested sampling algorithm.[6] The end products of the LALInference analyses are posterior samples for all of the parameters.

With parameter estimation, we appear to have another clear case of theory- or model-laden observation. In this case (unlike the case of template-based searches for gravitational waves) there is no alternative to model-based inference. All such "observations" of compact binary mergers are based on what the theory has to say about the behaviour of such systems.

5.2.3   Modelling Black Hole Coalescence

Overall, we have seen that both the detection of gravitational waves, through matched filtering, and the observation of compact binaries, through Bayesian parameter estimation techniques, rely on having accurate models of these phenomena.

It was known in advance that the best candidates for detection by the LIGO and Virgo interferometers were gravitational waves produced by compact binary mergers. These include binary systems containing two black holes, two neutron stars, or one of each. It was thus important to determine what general relativity predicted about the behavior of such systems, including their emission of gravitational waves. However, the general relativistic two-body problem is notoriously hard and we do not have exact analytic solutions for spacetimes containing merging compact binaries.[7] Thus we must rely on approximations, models, and simulations of these spacetimes in order to predict the form that any resultant gravitational waves will take. These include (1) post-Newtonian approximations, (2) models generated using black hole perturbation theory, (3) numerical relativity simulations, and (4) models based on the effective one-body approach.

---

[6] For details about LALInference analyses, see Veitch et al. (2015).

[7] See, for example, Kennefick (2007, Chap. 8) and Havas (1989, 1993) for the history of this problem.

(1) Post-Newtonian ("PN") theory is a well-established method for modelling systems where motions are slow compared to the speed of light and the gravitational field is weak.[8] PN models take Newtonian solutions as the starting point, then add general relativistic corrections to the equations of motion and the radiation field order by order in powers of v^2/c^2 (where v is the characteristic orbital velocity of the compact binary). Several different approaches to PN theory have been explored with the result that the PN equations of motion for two spinless black holes are now known up to 4PN order.[9] The gravitational radiation can be extracted through the multipolar post-Minkowskian wave generation formalism (Blanchet and Damour 1986; Blanchet 1987, 1998) or through the "direct integration of the relaxed Einstein equation" approach (Pati and Will 2000; Will and Alan 1996). PN theory is thought to model a compact binary and the resulting radiation well for the early inspiral. However, it fails to accurately model the late inspiral onwards, as the separation between the compact objects decreases and the orbital velocity becomes large.

(2) Black hole perturbation theory ("BHP") is a useful tool for modelling compact binaries with extreme mass ratios, where $m_2/m_1 \ll 1$.[10] For such mass ratios, the motion of the smaller object can be modelled by introducing perturbations in the background metric of the larger object. Leading order gravitational wave emission (related to the dissipative component of the self-force) is described by the Regge-Wheeler-Zerilli equations for a Schwarzschild background, and by the Teukolsky equation for a Kerr background.

(3) Numerical Relativity ("NR") waveform models are models produced through simulations on supercomputers. These simulations solve the exact Einstein equations numerically, enabling the calculation of the form of the gravitational waveform that would emanate from a binary merger with specific parameter values. After several decades of work to overcome a range of formidable technical challenges (e.g., formulations, gauge conditions, stable evolutions, black hole excision, boundary conditions, and wave extraction), numerical relativity saw a major breakthrough in 2005, with the first stable simulations of the final orbits, plunge, merger, and ringdown of a binary black hole merger. These original successes are reported in Pretorius (2005), Baker et al. (2006), and Campanelli et al. (2006).[11] Numerical relativity simulations are thought to be extremely accurate, even for the plunge and merger phases.[12] However, they are extremely computationally expensive, so it is not feasible to generate a 250,000-waveform template library from NR simulations alone.

(4) The effective one-body ("EOB") models of binary black holes are quasi-analytic models produced using the effective one-body formalism. This approach, developed by Alessandra Buonanno and Thibault Damour in the late 1990s, recasts the two-body problem as an effective field theory for a single particle (see Buonanno

---

[8] For a detailed discussion of the post-Newtonian approach, see Maggiore (2008, Chap. 5).

[9] Since the nPN order refers to inclusions of terms $\mathcal{O}(1/c^{2n})$, 4PN results include terms $\mathcal{O}(1/c^8)$.

[10] Here, by convention, $m_1 > m_2$.

[11] For detailed reviews of the history of numerical relativity simulations, see e.g., Holst et al. (2016) and Sperhake (2015).

[12] I discuss reasons for this confidence in Sect. 5.3.3.

and Damour (1999) and Buonanno and Damour (2000)). The basic idea behind this approach is to transform the conservative dynamics of two compact objects (masses m_1 and m_2, spins S_1 and S_2) into the dynamics of an effective particle ("mass" $\mu = m_1 m_2/(m_1 + m_2)$, and "spin" S^*) moving in a deformed Kerr metric ($M = m_1 + m_2$, S_{Kerr}). The EOB models build on the PN approach (by taking high-order PN results as input) in order to produce accurate analytic results for the entire process.

Each of these modelling approaches have different (but overlapping) domains of application. As mentioned above, PN models are valid for the early inspiral, but become increasingly inaccurate as the compactness parameter increases.[13] For BHP theory, the limiting factor is the mass ratio; the approximation becomes inaccurate as the mass ratio increases.[14] The domains of validity for PN and BHP do not have sharp cutoffs. Rather, the cutoff used depends on the acceptable level of error for a given calculation (Le Tiec 2014, 4). In principle, the domain of applicability for NR spans the whole parameter space. However, in practice, NR simulations are limited by available computing resources, with simulations involving large separations r or large mass ratios m_2/m_1 involving long and therefore costly calculations. In contrast, the EOB approach is supposed to be valid throughout the parameter space without being computationally expensive.

The models derived from each of these approaches are related in a range of ways. For example, EOB models and NR simulations take PN approximations as input; EOB models are calibrated ("tuned") against NR simulations; NR simulations are tested against PN approximations; and (calibrated) EOB models are tested against NR simulations.

In fact, the waveform models that are actually used by the LIGO-Virgo collaboration combine PN, EOB, and NR results. Two important families of hybrid models are the "EOBNR" and "IMRPhenom" models (Abbott et al. 2016c, 5). The EOBNR models are EOB models whose (unknown) higher order PN terms have been tuned to NR results in order to improve the Hamiltonian. They are thus hybrids in the sense that they are EOB models that are tuned to incorporate insights from both PN and NR results. Essentially, the EOB models have free parameters that can be modelled by comparison to NR simulations, producing the EOBNR models. The IMRPhenom models are hybrids in a more direct sense, in that they fit together inspiral, merger, and ringdown models derived from the PN, EOB, and NR approaches. These models are constructed by extending frequency-domain PN results then creating hybrids of PN and EOB models with NR waveforms. In particular, those used for parameter estimation in Abbott et al. (2016c, 5) are made by fitting untuned EOB waveforms and NR simulations.

---

[13] Here the compactness parameter is defined as as the ratio M/r where $0 < M/r \lesssim 1$, and $M = m_1 + m_2$.

[14] The mass ratio is defined as m_2/m_1 where, by convention, m_1 is always the larger mass and thus $0 < m_2/m_1 \lesssim 1$.

5.2.4   Model-Dependent Observation

The LIGO-Virgo Collaboration described GW150914 as the first "direct detection" of gravitational waves and the first "direct observation" of a binary black hole merger.[15] However, these descriptors have the potential to obscure the fact that these observations are also *indirect* in the sense that they are mediated by models, such as those in the EOBNR and IMRPhenom modelling families. Generating models that act as "mediating instruments" (Morgan and Morrison 1999) between theory and world was a vital contribution to the theoretical and technological advances that made the LIGO-Virgo experiments possible. As we have seen, "observations" in gravitational-wave astrophysics depend on the availability of these models.

The EOBNR and IMRPhenom models are a hodgepodge of other modelling approaches. Far from being a problem, this is what enables them to successfully bridge the gap between theory and data in the context of the LIGO-Virgo experiments. On the one hand, post-Newtonian approximation has a long history of bringing the theory of general relativity into empirical contact with observed systems, including the good agreement with observations of the Hulse-Taylor pulsars to high PN order. On the other hand, numerical relativity simulations provide the closest contact with the full Einstein field equations, since these simulations are as close as we have to exact solutions for the systems of interest. Thus determining what the theory of relativity tells us about the world is not a matter of analytically solving a system of differential equations. Rather, it involves heterogeneous modelling in a middle ground between theory and data, constrained by interfaces with existing theoretical and empirical results.

The EOBNR and IMRPhenom modelling frameworks are both thought to accurately describe gravitational wave signals for systems like the source of GW150914.[16] However, justifying confidence in these models, especially for the late stages of a binary black hole merger, is far from straightforward. This is largely due to the lack of either analytic solutions or previous empirical constraints in this regime. In the remainder of this paper, I critically examine the justification for confidence in models based on these approaches—both as faithful predictions derived from the Einstein field equations, and as accurate descriptions of real physical systems.

---

[15] In other work, including Elder (2020, in preparation), I provide an account of what is meant by these descriptors, drawing connections to recent work in the philosophy of measurement (e.g., Parker 2017; Tal 2012, 2013).

[16] For the purposes of this paper, I will mainly focus on the models used for this initial detection. However, modelling these systems is an active area of research, with new iterations of these modelling approaches incorporating more physical effects.

5.3   Model Validation and Circularity

When we turn to considering the justification for the models employed in the LIGO-Virgo experiments, the theory-ladenness or model-dependence of the observations becomes a source of potential concern. As with other instances of theory-ladenness, the general concern here is that any inaccuracy in the models could be systematically biasing observations and hence conclusions about the world. In Sects. 5.3.1 and 5.3.2, I examine how this general concern plays out for the observation of gravitational waves and binary black hole mergers, respectively.[17]

5.3.1   Worry 1: Theory-Laden Observation of Gravitational Waves

The use of matched filtering in the observation of gravitational waves appears to be a clear case of theory-laden observation, since this technique relies on advance knowledge of what the form of the gravitational wave signal will be. Traditional concerns about the theory-ladenness of observation are to do with more traditional notions of observation—the observations made by individual people—and how these are influenced by the conceptual schemes or "paradigms" of those individuals. Here the concern applies to a broader notion of observation that encompasses the out-puts of experimental procedures. Franklin (2015) calls this the "theory-ladenness of experiment". The observation of gravitational waves via matched filtering offers a particularly clear case of theory-ladenness. However, theory-ladenness also appears to be a generic feature of experiment. With this in mind, it might seem that the reliance on models in making observations can hardly be taken to be a special concern for LIGO-Virgo, but rather a general concern about all experiment-based observations. It is beyond the scope of this paper to address broader concerns about the theory-ladenness of either observations or experiments in general. (Indeed, I am inclined to doubt that such a general treatment would offer much insight into the particular concerns I address here.) Instead, my focus will be on the details of the present case and three potential problems that could arise due to the model-dependence of matched filtering: (1) missed signals, (2) false signals, and (3) sub-optimal signal extraction.

First, we might worry about gravitational wave signals being missed due to a lack of any corresponding templates. Such signals might lack corresponding templates because they originate from different types of sources than those represented in the template library, or because they exhibit behavior that deviates from the predictions of those templates (e.g., due to deviations from general relativity in the strong field regime). In either case, such signals would be a very interesting discovery and failure

---

[17] My focus here is on binary black hole mergers, in part because I am focused on the case of GW150914 specifically, and in part because the case is arguably slightly different for binary neutron star mergers. I discuss how the epistemic situation is changed for "multi-messenger" sources like GW170817 in Elder (2020, Chap. 4 and in preparation).

to detect them could be a major scientific loss. Missing such signals could also lead to a biased sampling of the kinds of events that produce gravitational waves detectable by LIGO and Virgo, leading to inaccurate conclusions about the populations of such events. However, there are some considerations that cut against this worry. For one thing, it is worth noting that matched filtering using general relativity-based templates should still be able to detect gravitational waves from a binary black hole merger unless the deviations from general relativistic descriptions are quite large (Yunes and Pretorius 2009, 1–2). It is also worth remembering that the data does not just disappear. This means data can be searched later. Indeed, work done by Abedi et al. (2017) to detect "echoes" in the LIGO-Virgo data provides an example of such a search. Finally, and perhaps most importantly, the LIGO-Virgo Collaboration also have unmodelled ("burst") search pipelines, which are able to detect gravitational wave signals with minimal assumptions about the form of the signal (Abbott et al. 2016b). These are less effective than the matched filtering search pipelines, reporting lower SNR and correspondingly lower statistical significance for the same gravitational wave signal. Furthermore, there have not been any confirmed detections from the unmodelled searches that were not also detected with modelled searches.[18] This makes it hard to determine the extent to which the unmodelled search pipelines help alleviate the concern that signals are being missed. However, the fact that these unmodelled searches are sensitive enough to detect the same events as the modelled searches (even if with lower SNR) at least demonstrates a reduced reliance on matched filtering for successful detection of gravitational waves.

Second, we might worry about whether the matched filtering procedure ever recovers false signals. After all, it seems that if we search for enough templates, through enough data, it is only a matter of time before the noise just happens to correlate with a template with a reported SNR above the designated threshold. If such false signals were being interpreted as genuine gravitational wave signals, there would indeed be something problematic about the data analysis procedures. However, this is the kind of problem that the LIGO methodology is best designed to avoid. Indeed, BICEP2's retraction was fresh in the minds of the LIGO-Virgo scientists, leading to an even higher level of caution about any detection claims (Collins 2017, 72).[19] The standards for detection implemented by the LIGO-Virgo Collaboration prioritize the avoidance of such false-positives, reflecting a high level of caution about any discovery claims. For example, a high SNR in one detector is not sufficient for an event to be classified as a detection. The LIGO-Virgo detection procedure also requires that there be coincident events in at least two detectors. For the modelled searches, these must be coincident triggers with the same template. Such a coincidence is unlikely to be mimicked by noise. Indeed, the LIGO-Virgo methods for analyzing the significance of the detection, using time slides, are designed to quantify the likelihood

---

[18] O3 has recently reported one burst candidate, given the preliminary name "S200114f", but this candidate has not yet been confirmed.

[19] "BICEP" stands for "Background Imaging of Cosmic Extragalactic Polarization". They initially reported the observation of signatures of primordial gravitational waves in 2014 but were forced to retract this claim, instead attributing the observation to cosmic dust (Collins 2017, 72).

that this coincidence could have been generated by chance due to detector noise. For GW150914, the false alarm rate was found to be less than one event per 203,000 years. In other words, it is thought to be extremely unlikely that the events detected were products of processes that contribute to the noise background.

Third, there is a kind of intermediate worry to be considered: the matched filtering procedure could extract a signal, but do so in such a way that the measured signal does not accurately represent the gravitational waves. This doesn't present a problem when it comes to detection, except perhaps in cases where the SNR is near the threshold value and imperfect extraction could be the difference between an event being classified as a detection or not. However, imperfect extraction of a signal does have implications for the inferences that we wish to make about it. For example, underestimating the amplitude of the wave could lead us to false conclusions about the distance to the compact binary, or the inclination of the orbital plane relative to the line of sight. More generally, the possibility of imperfectly filtered signals leads to worries about parameter estimation, and the further inferences we make based on biased "observations" of the compact binary systems. I turn to these issues in the following section.

5.3.2 Worry 2: Theory-Laden Observation of Binary Black Hole Mergers

The theory-ladenness, or model-dependence, of the LIGO-Virgo methodology becomes more problematic once we consider how the interferometers are used to "observe" compact binary coalescences. In particular, there is an apparently problematic circularity in the validation of these models, with epistemic implications for parameter estimation, theory testing, and further inferences about the population of such events. In this section, I will try to answer two important questions: First, what justification do we have for thinking that the models used by LIGO are good ones?; And second, what are the potential consequences of inaccuracies in these models?

There are two main sources of error to consider here: error due to modelling practices (PN, EOB, NR, etc.) and error due to the underlying theory (GR). Yunes and Pretorius (2009) classify these as "modelling bias" and "fundamental bias," respectively. More generally, we can characterise these sources of bias as being due to the inaccuracy of the model with respect to the underlying theory, in the case of modelling bias, and the inaccuracy of the theory with respect to the target system, in the case of fundamental bias. Some related distinctions are made elsewhere, with respect to the validity of models and simulations.

Verification and validation are prominent notions in the philosophical literature concerning computer simulations. These notions are derived from the corresponding scientific literature, since practitioners—especially in engineering and climate science contexts—have produced a large body of work dealing with these notions

(Winsberg 2019). Verification and validation are two categories of processes for checking the accuracy or reliability of a simulation.

Verification is the process of determining whether the simulation output is a good approximation to the actual solution of the equations of the relevant model. This can be broken up into two parts: checking that the computer code successfully implements the intended algorithm ("code verification") and checking that the outputs of this algorithm approximate the (analytic) solutions to the equations being solved via simulation ("solution verification") (Winsberg 2019). The latter, solution verification, sometimes takes the form of comparing simulation outputs to known analytic solutions that act as "benchmarks." In cases where no known benchmarks are on offer (as in the case of compact binary mergers) different approaches are needed. For example, it can be checked whether successive simulations converge on a stable solution as the resolution of the simulation is increased. In other words, this checks whether the result of the simulation stays approximately the same as the physical situation is modelled in an increasingly fine-grained manner.

Validation is the process of checking whether the given model is a good enough representation of the target system. Thus validation procedures require empirical tests of whether the simulation outputs provide a good enough description of the behaviour of the physical system of interest.

Verification concerns modelling bias. In the LIGO-Virgo case, verification procedures would be those processes designed to ensure that the outputs of numerical relativity simulations were true solutions to the Einstein field equations. Theoretical bias of the kind described by Yunes and Pretorius (2009) comes into play in the context of validation, since it is here that we are trying to establish whether the outputs of a simulation are accurate with respect to the physical world. Validation procedures must contend with any theoretical bias introduced into the broader testing procedure due the reliance on models that assume the accuracy of general relativity.

Some related concepts are those of internal and external validity.[20] The classic definitions of these notions are from Campbell and Staley (1966) and Cook (1979). According to these authors, *internal validity* concerns the validity of an inference to there being some kind of causal relationship that is captured by the experimental data. *External validity* concerns the validity of inferences about the generalisability of this causal relationship. In other words, an experiment is internally valid if we can gain genuine knowledge about the experimental system ("object") based on the data and it is externally valid if this knowledge allows us to make justified inferences about a system or class of systems that we want to learn about ("target").

Internal validity and external validity are related to modelling bias and fundamental theoretical bias through the roles models play in making inferences about both the object and target systems. Experiments involve a hierarchy of models that go between the theory and the experimental data (Suppes 1962). Morgan and Morrison (1999) discuss the many roles that models play in mediating between theory and data, arguing that they can be understood as technologies for learning—both about the theory and

---

[20] Other notions of validity can be added here. For example, see Cronbach and Meehl (1955) for a discussion of the notion of "construct validity".

about the world. Thus, for example, we use template waveforms as a tool for learning about the signal passing through the object system (through matched-filtering, as discussed above). We can also use these models for learning about the binary black hole system that produced the signal using sophisticated Bayesian inference software. Establishing internal and external validity in this case involves a series of model-based inferences for which both modelling bias and fundamental theoretical bias need to be taken into account.

5.3.3 Modelling Bias

Even if we can assume that general relativity provides an accurate description of systems like binary black hole mergers (an assumption I discuss below), there may still be reason to be concerned that the models used in detecting and reasoning about these systems are inaccurate. Given that we lack full analytic solutions against which to check our models, it is worth considering how these models are justified. In particular, how could we have enough confidence in these models that we were willing to rely on them in gravitational-wave astrophysics?

5.3.3.1 Validating EOBNR

As discussed above, PN, BHP, and NR have limited domains of applicability (though for different reasons). In contrast, the EOB formalism is supposed to offer a single framework that generates valid models throughout the parameter space. However, untuned EOB models are inadequate; it is EOBNR models, which have been tuned to NR simulations, that are needed for gravitational wave detection and compact binary parameter estimation.[21]

The EOBNR models are supposed to incorporate the results of the PN, EOB, NR and BHP approaches. Thus while the EOB models have some original physical motivation, confidence in the validity of these models is largely derived from confidence in other modelling approaches that are considered more secure. In particular, our confidence in these models in the strong dynamical regime (where PN models are no longer valid) is derived from our confidence in NR simulations.

NR simulations are believed to be extremely accurate. Sometimes these simulations are even referred to as "solutions" of the full Einstein field equations. For example, Abbott et al. (2016c, 3), states:

---

[21] For the purposes of this paper I will not explicitly discuss the IMRPhenom models. However, given that these hybrid models use the same ingredients as EOBNR models (PN, EOB, and NR), most if not all of what I say about EOBNR should also apply to IMRPhenom.

As the BHs get closer to each other and their velocities increase, the accuracy of the PN expansion degrades, and eventually the full solution of Einstein's equations is needed to accurately describe the binary evolution. This is accomplished using numerical relativity (NR) which, after the initial breakthrough, has been improved continuously to achieve the sophistication of modelling needed for our purposes.

What this means is that NR simulations are thought to be well verified with respect to general relativity. As with other simulations, this involves both code verification and solution verification.

However, solution verification is challenging in this context due to the lack of any full analytic solutions to use as benchmarks. Indeed, NR simulations are needed to act as benchmarks for other modelling approaches. To some degree, consistency with other modelling approaches—within their domains of applicability—can play a similar role. However, what is really needed is a way to verify NR simulations in the strong dynamical regime, where PN results are no longer adequate. Two important methods are used: convergence studies, and code comparison studies.

Convergence studies involve successive simulations of the same system with greater and greater resolution.[22] This is a valuable tool for establishing the accuracy of a simulation—in particular, for establishing that the result does not depend on the limited resolution used. If a study can show convergence to a stable solution, it can be reasonably inferred that this solution would remain valid if the resolution were increased. In other words, convergence studies are used to demonstrate that the results of a simulation are not an artifact of course-graining. In one of the breakthrough papers, Campanelli et al. (2006) perform such a study, demonstrating a high level of stability in the solution as the resolution is increased. Since NR simulations are extremely computationally expensive, it is important to avoid wasting computational resources when the gain in accuracy is insignificant. Developing an understanding of the relationships between resolution, convergence, and hence accuracy is thus important for practical reasons as well as for verification purposes.

Solution verification is also achieved by comparing the results from simulations conducted using different codes. It is worth noting that some of these codes employ very different methods. For example, the three breakthrough simulations of 2005 used two different approaches for dealing with the interiors of the black holes and the attendant singularities. Pretorius used a process called *excision*, according to which the black hole interior is simply not involved (and it need not be, since the region inside the event horizon cannot affect the region outside of it). The other two simulations used what has come to be called the *moving puncture* approach, according to which singularities are allowed to develop in the interior region, but are rendered sufficiently benign by an appropriate choice of gauge. The first code comparison was performed on codes used by each of these three groups: the LazEv code developed at Brownsville and Rochester Institute of Technology, the Hahndol code developed at NASA's Goddard Space Flight Center (GSFC), and Pretorius' original code. This study showed "exceptional agreement for the final burst of radiation, with some

---

[22] In general, convergence studies need not always involve increasing the resolution. For example, for Monte Carlo simulation convergence studies are performed for increasing sample sizes.

differences attributable to small spins on the black holes in one case" (Baker et al. 2007, Sect. 25).

Since then, more extensive comparisons have been performed. One important example is the Samurai project (Hannam et al. 2009) described below:

> For the Samurai project, comparisons were made between the SpEC ["Spectral Einstein Code"] code developed by the SXS ["Simulating eXtreme Spacetimes"] collaboration, the Hahndol code, the MayaKranc code developed at Penn State/Georgia Tech, the CCATI code developed at the Albert Einstein Institute, and the BAM ["Bi-functional Adaptive Mesh"] code developed at the University of Jena. One of the major differences between the Samurai project and [the earlier comparison] was the use of simulated LIGO noise data to determine if the differences between the waveforms generated by the various codes is, in practice, detectable.[23] (Duez and Zlochower 2018, 19)

Additional comparisons were conducted in the process of confirming the status of GW150914. The SXS Collaboration, which uses the SpeC code, and the Rochester Institute of Technology (RIT) group, which uses LazEv, compared the results of their attempts to model the source of GW150914 in Lovelace et al. (2016). The two codes used different initial data generation techniques, evolution techniques, and waveform extraction techniques, and shared no common routines. Despite these methodological differences, they found that the dominant modes produced by the two codes agreed to better than 99.9% (Duez and Zlochower 2018, 20). The level of agreement in comparisons such as this, combined with the simulators' confidence in their own codes, is the main source of confidence in the results of NR simulations of compact binary mergers.[24]

NR simulations of GW150914-like events are generally considered to be well-verified for the purposes of both detection and parameter estimation, despite the lack of analytic benchmarks. Indeed, the agreement between waveforms derived from independent simulations is such that the waveforms are indistinguishable below an SNR (Hannam et al. 2009).

However, they are too computationally expensive to generate all the templates needed for matched-filter-based detection. Instead, the NR simulations are used as a kind of benchmark for EOB models, which require minimal computational resources. The EOB models contain a number of free parameters that can be tuned to NR results. The entire parameter space can then be filled by interpolation between known NR results. While the EOB models do have some independent physical motivation, much of the confidence in templates based on this approach—that is, EOBNR models—is derived from confidence in NR.

---

[23] For more on these codes, see Duez and Zlochower (2018), especially section II, and the references therein).

[24] The agreement across variations in the simulation methods forms the basis for a robustness argument: the simulation outputs are considered to be robust (and hence reliable) due to the agreement across independent methods. As with other robustness arguments, such reasoning relies on the methods being genuinely independent (see e.g., Staley (2004) and Dethier (2020)). Note that while robustness arguments of this kind have been considered controversial in the context of climate modelling, the present case appears to be one where robustness arguments are considered to be uncontroversial (if fallible). However, a comparative study of these cases is beyond the scope of this paper.

The EOBNR models are effective mediating instruments for use in gravitational wave detection and compact binary observation. They combine insights from a range of modelling approaches, and are hence well verified through their relationships to these other models. However, producing models that incorporate all the relevant physical effects for the full range of possible binary systems is still a significant challenge. At the time of the O1 observing run, there were no models that could take account of all such effects (e.g., eccentricity and higher order modes in the presence of spin) for the full range of possible binary systems (Abbott et al. 2016c, 4). However, this is an area of active research; since GW150914, simulation studies have continued to explore new regions of the parameter space. For a summary of progress in numerical relativity simulations of compact binaries in the twenty-first century, see Duez and Zlochower (2018).

Ultimately, one major source of modelling error for the templates comes from the practical limitations on NR simulations. NR simulations can be used as benchmarks, to tune the EOB models, but they cannot be used exclusively to generate a whole template library. This means that the majority of templates have not been directly compared to full NR simulations (although they are based on extrapolations from such comparisons). Following detection and parameter estimation, new NR simulations are performed using the physical parameters that have been estimated from the source to test for agreement with the measured signal.

5.3.4   Theoretical Bias

Yunes and Pretorius (2009) discuss what they call a "fundamental bias" in the methodology of gravitational-wave astrophysics: the assumption of the validity of general relativity, and the Einstein field equations.

It is important to note at the outset that we do have high confidence in the theory of general relativity *within the regimes in which it has been tested*. So far, general relativity has stood up to every empirical test we have thrown at it, from Einstein's successful retrodiction of the precession of the perihelion of Mercury, to the prediction of the orbital decay of the Hulse-Taylor pulsars.[25] However, these tests alone cannot justify extrapolation to the extreme conditions present when two black holes coalesce.

The success of general relativity under previously-probed conditions provides no guarantee of its success under the extreme conditions present in a binary black hole merger. Binary black hole mergers involve both high velocities and strong gravity, placing them firmly in the dynamical strong field regime. For all we know, another theory of gravity (or quantum gravity) might distinguish itself as the better theory

---

[25] Of course, from the perspective of proponents of (relativistic extensions of) Modified Newtonian Dynamics (MOND), the empirical discrepancies that are usually attributed to dark matter in order to save general relativity are instead a motivation to modify general relativity. On this view, of course, general relativity has not stood up to all the empirical tests it has faced.

in such regimes. In advance of empirical investigations of such regimes, we cannot assume that general relativity provides an adequate description of merger dynamics.

These concerns about the theory in turn give some reason to be concerned about the descriptions of specific systems provided by our models of binary black hole mergers. After all, if the conditions present in binary black hole mergers turn out to be beyond the domain of applicability of general relativity, then any models based on this theory may also be inaccurate under these conditions. We do have good reason to think that the part of the waveform—the early inspiral—that is based on post-Newtonian approximations will hold up. After all, these approximations are within the regime that has been tested already by observation of Hulse-Taylor binaries. However, numerical relativity simulations of the plunge and merger may well be inaccurate if general relativity fails to be an accurate description of the system when we reach the dynamical strong field regime. Since these simulations are used as a benchmark for ensuring the accuracy of other models in the template bank, this inaccuracy would likely infect the other models too. Insofar as current models *are* good models of the system according to general relativity, any deviations of the dynamics of the system from those predicted by general relativity will render the models inaccurate descriptions of the target systems that they are supposed to represent.

The possibility of deviations from general relativity in the dynamical strong field regime leads to a fundamental bias in our inferences about these systems. This could have an impact both on the observation of gravitational waves (through matched-filtering) and on the inferences we make about the source (through parameter estimation).

First, theoretical bias could lead to non-optimal filtering. A possible example of this, given by Yunes and Pretorius (2009, 3), is that deviations from general relativity due to scalar radiation during late stages of the merger could lead to late time de-phasing with general relativity templates (due to inspiral occurring faster). This could lead to a systematically smaller SNR for detections, and thus systematic overestimation of the distance to the source. On a population level, we may end up concluding that such events occur more often farther away (i.e. further in the past).

Second, theoretical bias can be introduced at the level of parameter estimation, where the assumption of the accuracy of general relativity leads us to the inaccurate conclusions about the systems being observed. Yunes and Pretorius (2009, 3) provide the following example of this:

> For a second hypothetical example, consider an extreme mass ratio merger, where a small compact object spirals into a supermassive BH [black hole]. Suppose that a Chern-Simons (CS)-like correction is present, altering the near-horizon geometry of the BH [...] To leading order, the CS correction reduces the effective gravitomagnetic force exerted by the BH on the compact object; in other words, the GW emission would be similar to a compact object spiraling into a GR Kerr BH, but with smaller spin parameter a. Suppose further that near-extremal (a $\approx$ 1) BHs are common (how rapidly astrophysical BHs can spin is an interesting and open question). Observation of a population of CS-modified Kerr BHs using GR templates would systematically underestimate the BH spin, leading to the erroneous conclusion that near-extremal BHs are uncommon, which could further lead to incorrect inferences about astrophysical BH formation and growth mechanisms.

Thus the assumption that general relativity provides an accurate description of black hole coalescence may bias parameter estimation and any subsequent inferences.

5.4 Model Validation with the LIGO-Virgo Observations

We have now seen that there are a range of reasons that the models used by the LIGO-Virgo Collaboration might provide inaccurate descriptions of binary black hole mergers, at least in the late stages of these events. Any inaccuracies could lead to systematic biases in the inferences that we make about such systems. In advance of any empirical testing, it is impossible to be sure that these models are accurate. *Prima facie* it seems possible that the LIGO-Virgo measurements themselves could be used to validate the models of the systems that they are observing. Thus they could be used to demonstrate that general relativity provides a good description of such events. However, this is called into question by the model-dependence of the observations, in terms of both the matched filtering needed to optimally retrieve gravitational wave signals, and the Bayesian parameter estimation used to determine the properties of the system being observed.

The basic problem is this: testing the validity of the models using the LIGO-Virgo observations relies on a parameter estimation process that presupposes the validity of the models being used—models such as EOBNR and IMRPhenom. Thus any empirical test seems to implicate us in a circular justification scheme. We can only test the predictions of general relativity if we know the properties (mass, spin, etc.) of the objects we are observing, but we can only estimate these properties by assuming that our general relativistic models of these objects are accurate. Essentially, this is because we can only test general relativity insofar as we test its predictions about the dynamical behaviour of known objects (and consequences of this for gravitational wave emission, etc.). However, for binary black hole mergers, our only way of learning about these objects is via models that presuppose the accuracy of general relativity within the dynamical strong field regime.

Clearly this circularity problem has some connections with (Collins 1985)s "Experimenter's Regress," according to which we do not know whether we have made a good measuring device until we have one that gives us the right results, but we do not know what the right results are until we know that we have a good measuring device (that is producing those results). Aside from describing this problem as a regress, Collins sometimes describes this as a circularity between the measurement device and the measurement result; the validity of each depends on the validity of the other. According to Collins, using an experiment as a test requires finding a way 'to break into the circle' (Collins 1985, 84). Controversially, Collins thinks that this circularity is broken by social negotiation rather than rational arguments, while others, such as Franklin (1994) argue that epistemic criteria are sufficient to break the circle.

In the case of the circularity problem I have described for LIGO-Virgo, it is not (primarily) the detectors themselves that are in question. In this sense, it is a distinct problem to the one concerning Collins, which is best understood as applying to gravitational wave detection.[26] *Even if* we think that we can 'break the circle' described by Collins (i.e., we are confident that the interferometers have been successfully used to detect gravitational waves), the circularity I describe with respect to the observation of black holes presents a further problem to be resolved.

When justifying the LIGO-Virgo results qua detection of gravitational waves, confidence in the detector plays an important role (Elder 2020, in preparation). Here, the kinds of rational arguments described by Franklin feature prominently. For example, confidence in understanding the response of the detector is established through calibration procedures, using lasers to push the interferometer test masses, mimicking the effect of a passing gravitational wave. Without dismissing the social dimensions of scientific collaboration and discovery, I think it is fair to say that these epistemic considerations play a persuasive justificatory role in the the LIGO-Virgo detection claim. However, when it comes to the observation of binary black hole mergers, the circularity I have described presents a further (and more difficult) problem because these usual avenues for breaking the circle are unavailable.

Beyond the circularity itself, other features of the epistemic situation make the problem of model validation a particularly challenging one. First, the binary black hole systems being observed are distant astrophysical systems, meaning that it is not possible to manipulate or intervene on them in any way. Thus the analogue of calibration is not possible, since we cannot test the interferometer response to a binary black hole merger with known properties. Second, we have no independent access to these systems, given that black holes emit no electromagnetic radiation. This rules out using consilience, or coherence testing to improve confidence in the LIGO-Virgo results (and methods).[27] Third, the mergers themselves occur in regimes that have never been probed before—the dynamical strong field regime. This means that the previous success of general relativity provides no guarantee that the theory will continue to provide accurate descriptions in the regimes being probed by LIGO-Virgo. The lack of interventions or independent empirical access combined with the novel regimes being probed renders the problem of theoretical bias particularly acute, bringing the circularity problem beyond more generic issues of theory- or model-ladenness in empirical science. This circularity threatens to mask any bias implicit in the models.

Nonetheless, the LIGO-Virgo Collaboration does perform a number of tests of general relativity. Such tests seem to offer some empirical validation of the general relativity-based models used to observe binary black hole mergers. Indeed, taken together, they seem to constitute a kind of methodological bootstrapping (Glymour 1975). Each of the tests probes different assumptions that go into the observation of binary black hole mergers—despite making some assumptions based on general

---

[26] Nonetheless, insofar as the models are embedded in the experimental methodology, there is a sense in which this could be understood as a kind of experimenter's regress, spelled out in terms of models rather than detectors. There are also clear similarities here to the related "simulationist's regress" (Gelfert 2012; Meskhidze 2017).

[27] See Bokulich (2020) for discussion of the distinction between consilience and coherence testing.

relativity in the course of the tests. In particular, these tests involve looking for evidence that the model-dependent methodology—and circularity—could be biasing observations.

A detailed examination of theory-testing with LIGO-Virgo will be the subject of a future paper. However, a brief consideration of two tests helps illustrate how the LIGO-Virgo Collaboration is able to empirically validate models like EOBNR in the face of the circularity problem.[28]

First, the "residuals test" considered in Abbott et al. (2016d) tests the consistency of the residual data with noise. This involves subtracting the best-fit waveform from the GW150914 data and then comparing the residual with detector noise (for time periods where no gravitational waves have been detected). The idea here is to check whether the waveform has successfully removed the entire gravitational-wave signal from the data, or whether some of the signal remains. This process places constraints on the residual signal, and hence on the deviations from the best-fit waveform that could be present in the data. However, this doesn't constrain deviations from general relativity *simpliciter*, due to the possibility that the best-fit general relativity waveform is degenerate with non-general relativity waveforms for events characterized by different parameters. That is, the same waveform could be generated by a compact binary merger (described by parameters different from those that we think describe the GW150914 merger) with dynamics that deviate from general relativistic dynamics. In this case, we could be looking at different compact objects than we think we are, behaving differently than we think they are, but nonetheless producing very similar gravitational wave signatures. This is stated (though not fully explained) in the following passage:

> We use this estimated level [of residual] to bound GR violations *which are not degenerate with changes in the parameters of the binary* (Abbott et al. 2016d, 2, emphasis mine).

This test could potentially show inconsistency with general relativity, but not all deviations from general relativity will be detectable in this way. Thus the test shows that GW150914 is consistent with general relativity, but the methodology of this test could also be masking such deviations due to the fundamental bias associated with assuming general relativity for the purposes of parameter estimation.

Second, the "IMR consistency test" considered in Abbott et al. (2016d) considers the consistency of the (early) low-frequency part of the gravitational-wave signal with the (later) high-frequency part. This test proceeds as follows. First, the masses and spins of the two compact objects are estimated from the inspiral (low-frequency), using LALInference. This gives posterior distributions for component masses and spins. Then, using formulas derived from numerical relativity, posterior distributions for the remnant, post-merger object are computed. Finally, posterior distributions are also calculated directly from the measured post-inspiral (high-frequency) signal, and the two distributions are compared. These are also compared to the posterior distributions computed from the inspiral-merger-ringdown waveform as a whole.

---

[28] See also Patton (2020) for an excellent discussion of a different test, based on the parameterised post-Einsteinian framework developed by Yunes and Pretorius, including how this connects to the issue of "fundamental theoretical bias".

If there are any deviations from general relativity to be found, these are expected to occur in the late part of the signal, where the full non-linear Einstein field equations are needed and approximations are known to become invalid. In contrast, previous empirical constraints give us reason to doubt that such deviations will be significant for the early inspiral. In the presence of high-frequency deviations, parameter estimation based on general relativity models will deviate from the values of a system that is well-described by general relativity. Hence (in such cases) we can expect the parameter values estimated from the low frequency part of the signal to show discrepancies with the parameter values estimated from the high frequency part of the signal. This leaves the test open to detection of subtle deviations from general relativity; if the parameters associated with the two waveforms are different, this could suggest some deviation from general relativity (Abbott et al. 2016d). Interestingly, it does so despite assuming the validity of general-relativistic descriptions at each step in the process.

Although these two tests are just consistency tests, they place constraints on ways that the errors in the models could be undermining the accuracy of observations. Taken together, they constrain both the accuracy of the extracted gravitational waveform and the consistency of this waveform with general relativity. Further tests performed by the LIGO-Virgo place further constraints on loopholes in their methods—that is to say, these tests place constraints on the extent to which particular aspects of the LIGO-Virgo methodology could be biased by inadequate modelling of the target system. In doing so, they can be understood as a response to the circularity problem that I have described in this paper.[29]

5.5 Conclusion

In this paper, I have argued that modelling plays an essential role in connecting high-level theory, embodied in the Einstein field equations, with the LIGO-Virgo data. The models used in template-based searches for gravitational waves and in parameter estimation incorporate insights from a range of modelling approaches, allowing us to gain empirical access to binary black hole mergers.

However, I have also argued that the model-dependent methods used by the LIGO-Virgo Collaboration to observe binary black hole mergers lead to some epistemic challenges; the potential bias introduced through the use of general relativity-based models leads to a circularity problem for the validation of these models in the regimes probed by the LIGO-Virgo Collaboration. Observations of binary black hole systems are based on models of such systems, and confidence in the accuracy of these observations depends on the validity of the models being used. Thus using these observations to validate these models is problematically circular. (However, I briefly mentioned

---

[29] For further discussion of theory testing in gravitational-wave astrophysics, see Elder (in preparation).

some ways that the LIGO-Virgo proceeds in validating their models in spite of this circularity.)

Overall, this paper shows how the methodology of the LIGO-Virgo experiments is intimately bound up with models—of binary black hole mergers and the gravitational waves that they produce. The success of these experiments rests on confidence in these models, which bridge the gap between theory and phenomena. The flip side of this is that much of the interesting, and challenging work in validating the LIGO-Virgo results lies in validating the models themselves with respect to both the equations of general relativity and the physical systems being observed.

Acknowledgements I would like to thank Don Howard, Nicholas Teh, Feraz Azhar, and Erik Curiel for their support and comments on an early version of this paper. I would similarly like to thank Dennis Lehmkuhl and the rest of the Lichtenberg Group at the University of Bonn—especially Juliusz Doboszewski, Niels Martens, and Christian Röken—for their helpful feedback on early drafts.
I would also like to thank audiences at SuperPAC, MS8, EPSA, the BHI and the (history and) philosophy of physics seminars at Bonn and Oxford for their questions and comments. Conversations with members of the LIGO Scientific Collaboration (especially Daniel Holz and Hsin-Yu Chen) at Seven Pines, BHI conferences, and a Sackler conference were also invaluable.
Special thanks are also due to Lydia Patton, a pioneer in the philosophy of gravitational-wave astrophysics, for ongoing support of my work in this area. And, of course, for thinking to connect this paper with the other excellent work in this volume.
This project/publication was funded in part by the Gordon and Betty Moore Foundation. It was also made possible through the support of a grant from the John Templeton Foundation. The opinions expressed in this publication are those of the author and do not necessarily reflect the views of these Foundations.

References

Abbott BP et al (2016a) Characterization of transient noise in advanced LIGO relevant to gravitational wave signal GW150914. Class Quantum Grav 33(13):134001. https://doi.org/10.1088/0264-9381/33/13/134001

Abbott BP et al (2016b) Observing Gravitational-Wave transient GW150914 with minimal assumptions. Phys Rev D 93(12):122004. https://doi.org/10.1103/PhysRevD.93.122004

Abbott BP et al (2016c) Properties of the binary black hole merger GW150914. Phys Rev Lett 116(24):241102. https://doi.org/10.1103/PhysRevLett.116

Abbott BP et al. (2016d) Tests of general relativity with GW150914. Phys Rev Lett 116(22):221101. https://doi.org/10.1103/PhysRevLett.116.221101

Abbott BP et al (2020) A guide to LIGO-Virgo detector noise and extraction of transient gravitational-wave signals. Class Quantum Grav 37(5):055002. https://doi.org/10.1088/1361-6382/ab685e

Abedi J, Dykaar H, Afshordi N (2017) Echoes from the Abyss: tentative evidence for Planck-scale structure at black hole horizons. Phys Rev D 96(8):082004

Baker JG, Campanelli M, Pretorius F, Zlochower Y (2007) Comparisons of binary black hole merger waveforms. Class Quantum Grav 24(12):S25–S31. https://doi.org/10.1088/0264-9381/24/12/s03

Baker JG, Centrella J, Choi D-I, Koppitz M, van Meter J (2006) Gravitational-Wave extraction from an inspiraling configuration of merging black holes. Phys Rev Lett 96(11). https://doi.org/10.1103/PhysRevLett.96.111102

Blanchet L, Damour T (1986) Radiative gravitational fields in general relativity I. General structure of the field outside the source. In: Philosophical transactions of the royal society of London. Series a, mathematical and physical sciences, vol 320(1555), pp 379–430

Blanchet L (1987) Radiative gravitation fields in general relativity II. Asymptotic behaviour at future null infinity. In: Proceedings of the royal society of London. Series a, mathematical and physical sciences, vol 409(1837), pp 383–399

Blanchet L (1998) On the multipole expansion of the gravitational field. Class Quantum Grav 15(7):1971–1999. https://doi.org/10.1088/0264-9381/15/7/013

Bokulich A (2020) Calibration, coherence, and consilience in radiometric measures of geologic time. Philos Sci 87(3):425–456

Buonanno A, Damour T (1999) Effective one-body approach to general relativistic two-body dynamics. Phys Rev D 59(8):084006. https://doi.org/10.1103/PhysRevD.59.084006

Buonanno A, Damour T (2000) Transition from inspiral to plunge in binary black hole coalescences. Phys Rev D 62(6): 064015. https://doi.org/10.1103/PhysRevD.62

Campanelli M, Lousto CO, Marronetti P, Zlochower Y (2006) Accurate evolutions of orbiting black-hole binaries without excision. Phys Rev Lett 96(11). https://doi.org/10.1103/PhysRevLett.96.111101

Campbell D, Staley J (1966) Experimental and quasi-experimental designs for research. R. McNally, Chicago

Collins H (1985) Changing order: replication and induction in scientific practice. University of Chicago Press

Collins H (2017) Gravity's kiss: the detection of gravitational waves. MIT Press, Cambridge, MA

Cook TD (1979) Quasi-experimentation: design and analysis issues for field settings. Houghton Mifflin, Boston

Cronbach L, Meehl P (1955) Construct validity in psychological tests. Psychol Bull (Washington, etc) 52

Dethier C (2020) Multiple models, robustness, and the epistemology of climate science. Notre Dame, Indiana

Duez MD, Zlochower Y (2018) Numerical relativity of compact binaries in the 21st century. Rep Prog Phys 82(1):016902. https://doi.org/10.1088/1361-6633/aadb16

Elder J (2020) The epistemology of gravitational-wave astrophysics. PhD diss, University of Notre Dame

Elder J (in preparation) Independent evidence in multi-messenger astrophysics

Elder J (in preparation) On the 'Direct Detection' of gravitational waves

Elder J (in preparation) Theory testing in gravitational-wave astrophysics

Fillion N, Bangu S (2015) Numerical methods, complexity, and epistemic hierarchies. Philos Sci 82(5):941–955

Franklin A (1994) How to avoid the experimenters' regress. Stud Hist Philos Sci Part A 25(3):463–491. https://doi.org/10.1016/0039-3681(94)90062-0. Black Hole Coalescence: Observation and Model Validation 25

Franklin A (2015) The theory-ladenness of experiment. J Gen Philos Sci 46(1):155–166. https://doi.org/10.1007/s10838-015-9285-9

Gelfert A (2012) Scientific models, simulation, and the experimenter'sregress. In: Imbert C, Humphreys P (eds) Models, simulations, and representations. Routledge studies in the philosophy of science, vol 9, pp 145–167. Routledge, New York

Glymour C (1975) Relevant evidence. J Philos 72(14):403–426

Hannam M, Husa S, Baker JG, Boyle M, Brügmann B, Chu T, Dorband N et al (2009) Samurai project: verifying the consistency of black-hole-binary waveforms for gravitational-wave detection. Phys Rev D 79 (8):084025. https://doi.org/10.1103/PhysRevD.79.084025

Havas P (1989) The early history of the problem of motion in general relativity. In: Howard D, Stachel J (eds) Einstein and the history of general relativity: based on the proceedings of the 1986 Osgood Hill Conference, North Andover, MA, 8–11 May 1986. Einstein Studies, vol 1. Birkhäuser, Boston

Havas P (1993) The two-body problem and the Einstein-Silberstein controversy. In: Earman J, Janssen M, Norton J (eds) The attraction of gravitation: new studies in the history of general relativity. Einstein studies, vol 5. Birkhäuser, Boston

Holst M, Sarbach O, Tiglio M, Vallisneri M (2016) The emergence of gravitational wave science: 100 years of development of mathematical theory, detectors, numerical algorithms, and data analysis tools. Bull Am Math Soc 53:513–554. https://doi.org/10.1090/bull/1544

Kennefick D (2007) Travelling at the speed of thought. Princeton University Press

Le Tiec A (2014) The overlap of numerical relativity, perturbation theory and post-Newtonian theory in the binary black hole problem. Int J Mod Phys D 23(10):1430022. https://doi.org/10.1142/S0218271814300225

Lovelace G, Lousto CO, Healy J, Scheel MA, Garcia A, O'Shaughnessy R, Boyle M et al (2016) Modeling the source of GW150914 with targeted numerical-relativity simulations. Class Quantum Grav 33(24):244002. https://doi.org/10.1088/0264-9381/33/24/244002

Maggiore M (2008) Gravitational waves. Theory and experiments, vol 1. Oxford University Press, Oxford

Meskhidze H (2017) Simulationist's Regress in Laboratory Astrophysics

Morgan M, Morrison M (1999) Models as mediating instruments. In: Morgan M, Morrison M (eds) Models as mediators: perspectives on natural and social science, pp 10–37. Cambridge University Press

Parker WS (2017) Computer simulation, measurement, and data assimilation. Br J Philos Sci 68(1):273–304. https://doi.org/10.1093/bjps/axv037

Pati ME, Will CM (2000) Post-Newtonian gravitational radiation and equations of motion via direct integration of the relaxed Einstein equations: foundations. Phys Rev D 62(12):124015. https://doi.org/10.1103/PhysRevD.62.124015

Patton L (2020) Expanding theory testing in general relativity: LIGO and parametrized theories. In: Studies in history and philosophy of science. Part B: studies in history and philosophy of modern physics, vol 69, pp 142–153. https://doi.org/10.1016/j.shpsb.2020.01.001

Peters PC (1964) Gravitational radiation and the motion of two point masses. Phys Rev 136(4B). https://doi.org/10.1103/PhysRev.136.B1224

Pretorius F (2005) Evolution of binary black-hole spacetimes. Phys Rev Lett 95(12). https://doi.org/10.1103/PhysRevLett.95.121101

Shapere D (1982) The concept of observation in science and philosophy. Philosop Sci 49(4). https://doi.org/10.1086/289075

Sperhake U (2015) The numerical relativity breakthrough for binary black holes. Class Quantum Grav 32(12):124011. https://doi.org/10.1088/0264-9381/32/12/124011

Staley KW (2004) Robust evidence and secure evidence claims. Philosop Sci 71(4):467–488. https://doi.org/10.1086/423748

Suppes P (1962) Models of data. In: Nagel E, Suppes P, Tarski A (eds) Logic, methodology and philosophy of science: proceedings of the 1960 international congress, pp 252–261. Stanford University Press

Tal E (2012) The epistemology of measurement: a model-based account. http://search.proquest.com/docview/1346194511/

Tal E (2013) Old and new problems in philosophy of measurement. Philosop Compass 8(12):1159–1173. https://doi.org/10.1111/phc3.12089

Veitch J et al (2015) Parameter estimation for compact binaries with groundbased gravitational-wave observations using the LAL inference software library. Phys Rev D 91(4):042003. https://doi.org/10.1103/PhysRevD.91.042003

Will CM, Alan GW (1996) Gravitational radiation from compact binary systems: gravitational waveforms and energy loss to second post-newtonian order. Phys Rev D 54(8):4813–4848. https://doi.org/10.1103/PhysRevD.54.4813

Winsberg E (2019) Computer simulations in science. In: Zalta EN (ed) The Stanford Encyclopedia of Philosophy, Spring 2019. Metaphysics Research Lab, Stanford University

Yunes N, Pretorius F (2009) Fundamental theoretical bias in gravitational wave astrophysics and the parametrized post-Einsteinian framework. Phys Rev D 80(12):122003. https://doi.org/10.1103/PhysRevD.80.122003